AUTUMN IN THE FOOTHILLS OF THE SAN JUAN MOUNTAINS

MIDSUMMER FLOES IN ICEBERG LAKE, GLACIER NATIONAL PARK

PEAKS OF THE ROCKIES, NORTHERN MONTANA

MULE DEER IN GLACIER NATIONAL PARK

ASPENS AND WILD FLOWERS IN WYOMING'S TETON NATIONAL FOREST

SUNSET AT OREGON BUTTES IN THE GREAT DIVIDE BASIN

MAROON LAKE AND THE PEAKS OF THE MAROON BELLS NEAR ASPEN, COLORADO

Other Publications:

THE GOOD COOK
THE SEAFARERS
THE ENCYCLOPEDIA OF COLLECTIBLES
WORLD WAR II
THE GREAT CITIES
HOME REPAIR AND IMPROVEMENT
THE WORLD'S WILD PLACES
THE TIME-LIFE LIBRARY OF BOATING
HUMAN BEHAVIOR
THE ART OF SEWING
THE OLD WEST
THE EMERGENCE OF MAN
THE TIME-LIFE ENCYCLOPEDIA OF GARDENING
LIFE LIBRARY OF PHOTOGRAPHY
THIS FABULOUS CENTURY
FOODS OF THE WORLD
TIME-LIFE LIBRARY OF AMERICA
TIME-LIFE LIBRARY OF ART
GREAT AGES OF MAN
LIFE SCIENCE LIBRARY
THE LIFE HISTORY OF THE UNITED STATES
TIME READING PROGRAM
LIFE NATURE LIBRARY
LIFE WORLD LIBRARY
FAMILY LIBRARY:
HOW THINGS WORK IN YOUR HOME
THE TIME-LIFE BOOK OF THE FAMILY CAR
THE TIME-LIFE FAMILY LEGAL GUIDE
THE TIME-LIFE BOOK OF FAMILY FINANCE

THE GREAT DIVIDE

THE AMERICAN WILDERNESS/TIME-LIFE BOOKS/ALEXANDRIA, VIRGINIA

BY BRYCE S. WALKER
AND THE EDITORS OF TIME-LIFE BOOKS

The Author: Long devoted to tramping through the Berkshires of his native Massachusetts, Bryce Walker became converted to the Rocky Mountains in the course of nearly a year spent climbing, riding and skiing in the Great Divide country. He has been a travel writer for United Press International, and was for several years a staff writer for TIME-LIFE BOOKS, contributing to volumes on science, photography, and American and world history.

The Cover: Starkly delineated by late afternoon sunlight, the Great Divide snakes along a snowy ridge that marks the site of the continental watershed in the San Juan Mountains of southwestern Colorado.

Third printing. Revised 1979.
Published simultaneously in Canada.
Library of Congress catalogue card number 73-81327.

Contents

Where a Continent's Waters Divide

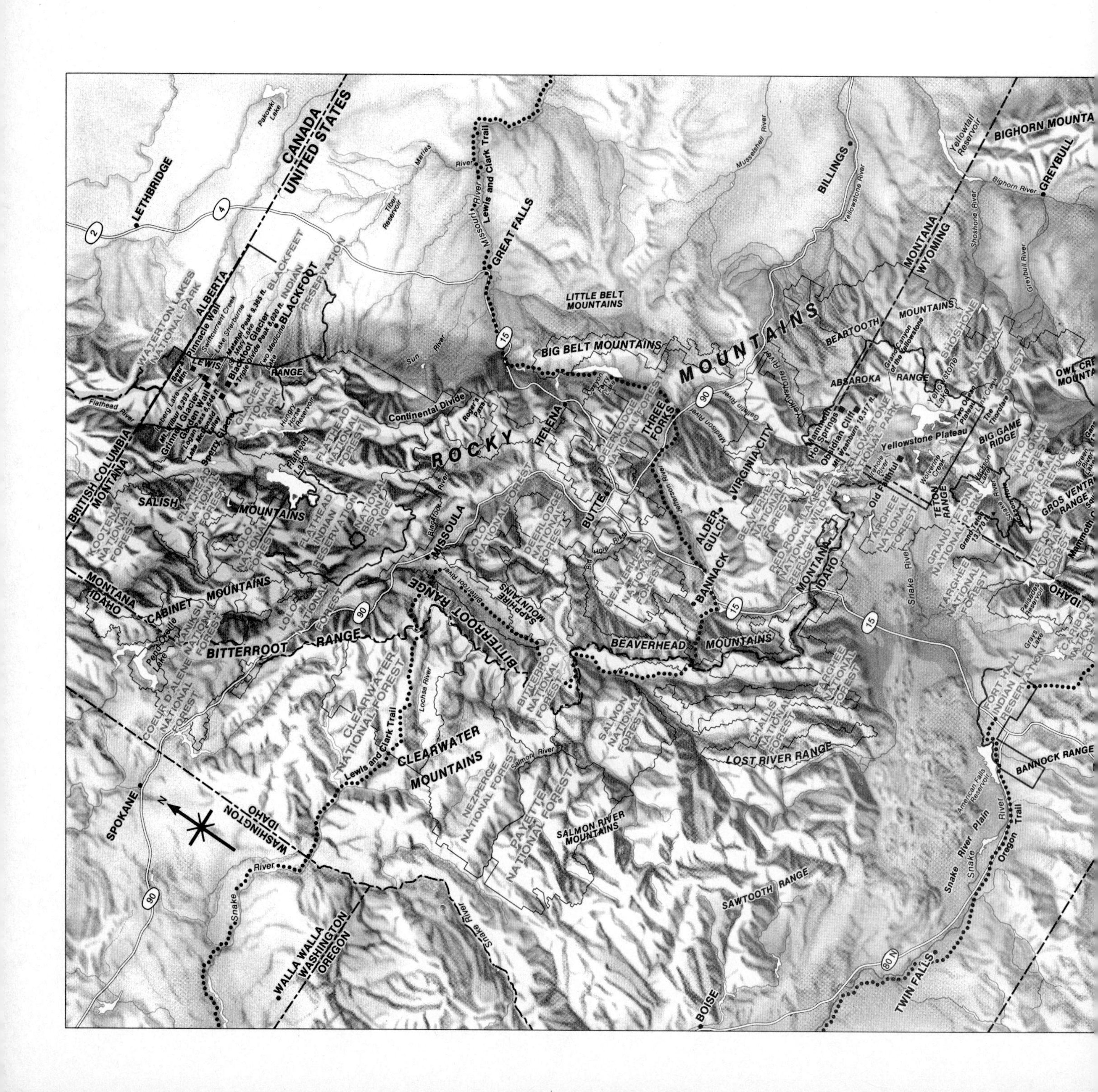

The Great Divide country explored in this book (detailed map, below) touches five states as it follows the continental watershed in its 1,700-mile course through the United States Rockies from northern New Mexico to the Canadian border. The divide is represented by a thick blue line. Terrain above 10,000 feet in the mountainous country that the divide traverses is indicated in white, terrain between 5,000 and 10,000 feet in yellow, and that below 5,000 feet in green. Rivers are shown as blue lines, lakes are white. Black triangles locate major peaks, black squares denote points of special interest. Thin red lines demark national forests, and heavier red lines enclose other federal areas such as national parks and wildlife refuges. Two historical routes heavily traveled by pioneers crossing the divide, the Oregon Trail and the Lewis and Clark Trail, are traced by black dotted lines.

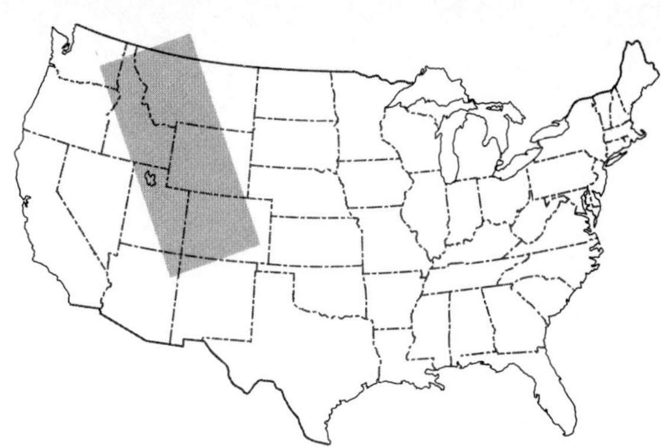

1/ The Roof of the Continent

We were in such an airy elevation above the creeping populations of the earth that...it seemed that we could look around and abroad and contemplate the whole great globe. MARK TWAIN/ *ROUGHING IT*

It is an astonishing concept. An immense cordon of mountain ranges, two continents in length, that splits the waters of the Western Hemisphere from Alaska to the Strait of Magellan. You cannot travel from Kansas to Nevada without crossing it, or from Chile to Argentina. You could fly down it from dawn to dusk in the world's fastest commercial airliner, the stupendous crags shrunk to slag heaps in the airplane window, and still not reach the end. To a man who stands on its crest in the Rocky Mountains of the United States, two miles higher than the level of the ocean, with the clean wind spilling into his face, the turbulence of landscape seems to go on forever—crags, steeples, ridges, canyons, rock slides, snow fields, battlements, glaciers, minarets to the utmost reaches of the imagination. This is the continent's spine, where the skeleton pokes through: the mountain barrier of the Great Divide.

No man will ever know all of the Great Divide. Its enormous variety —even that part of it within the United States—will not fit into the confines of a single brain. To a student of geography it is officially known as the Continental Divide. In the United States it is a line on a map that enters the country in northwestern Montana and winds roughly southward along the tumbled ranges of the Rocky Mountains, through Wyoming and Colorado, and exits via the New Mexico border. Others think of it differently: as a place to hunt elk, find gold, climb mountains, look for grizzly bears, study geology, predict the weather. A visitor to Glacier National Park in Montana can see the divide quite clearly; it looms a thousand feet overhead, a sheer rock wall thin and jagged as a saw blade. Yet it lies so flat across one portion of the Wyoming prairie that a man can scarcely tell when he has reached its crest. For a skier in Colorado the divide is a snowbank that arrives in October and piles 150 inches deep before melting the following May. For a man in search of solitude it represents the most prodigious swath of untamed land in the lower 48 states.

You may straddle the divide in a spruce forest, an aspen grove, a desert, a meadow or at the lip of a precipice that drops to eternity. You may follow it above tree line and across the tundra, where temperatures can be arctic and the flowers as rare and fragile as any on earth. But, in theory at least, one thing you will not do is get your feet wet. Not a single river crosses the divide, and all water flows away from it.

There is nothing extraordinary about most divides. They are simply high areas that separate the watersheds of rivers. Rain falling on the east face of a ridge drains into one river system, while rain falling on the west face drains into another. The ridge acts like the crown of a road, diverting water into the ditches on either side. There may be several such ridges in a single square mile of land; America has thousands of tiny divides, separating a multitude of Bear Creeks from Willow Creeks from Trout Forks from Piney Lakes. But nowhere in the world is there a divide of such imperial proportions—or one that parts the waters with such a grand, decisive gesture—as the Great Divide.

This titanic watershed separates not just rivers, but oceans. Here on the divide's east slope the Missouri River sets out to join the Mississippi and eventually to flow almost 4,000 miles to the Gulf of Mexico and so to the Atlantic. The Platte also, and the Arkansas, roll down the mountains, out across the plains, mingling their waters with those of the Gulf and the Atlantic Ocean. Down the west slope, toward the Pacific, plunge the Columbia, the Snake, the Colorado. You might, if your mind turns that way, hypothesize a snowflake falling on one slope of the divide. In the spring thaw it will run down a network of rills, creeks, sluices, cataracts and other waterways until it adds a few molecules of moisture to the ocean. Provided, of course, it does not seep into the ground, evaporate or get drunk by a mule deer or a pine tree first.

Most people first glimpse the Great Divide from the eastern, Atlantic side. It has always been so, ever since the days of the Conestoga wagons. The traveler follows the sun across the Great Plains, through sagebrush and alkali dust, the green tips of cottonwoods showing above the river bottoms. The air grows thin and crisp, the colors intensify until the line between shadow and sunlight is as sharp as the cut of a knife. The sky turns cobalt, with a brightness that seems almost artificial. A low line of clouds smudges the western horizon. Then the clouds separate, and their lower billows harden into rock: these are the mountains, with the divide weaving along their tops.

In many places the barrier rises from the plains with startling abruptness; along the Front Range of the Rockies in central Colorado, only a narrow band of foothills effects the transition. From a distance the plains simply seem to stop. They quit as abruptly as an automobile crashing into a concrete wall. Yet the plains themselves have been rising imperceptibly all the way from the Mississippi until, at a locale such as Boulder, Colorado, the land is a mile above sea level. Except for a few isolated peaks, this is higher than any other piece of land to the east for 5,000 miles—at which point you reach the Sierra de Gredos in Spain. You have yet to set foot on a Rocky Mountain. Behind Boulder the body of the Front Range soars upward another full mile, and the peaks of the divide another half mile above that. They seem very close, and they are—close enough to hit with a long-range artillery shell.

But generally hard to get to. Despite more than a century of reasonably civilized habitation, of town building and road making, of mining and timber cutting and cattle ranching, the region of the Great Divide is still one of the least populated in the United States. Only 30 paved roads cross the divide between Canada and New Mexico, and some of these are closed in winter. Silverton, in southern Colorado, is only 40 miles as the crow flies from Creede, another town on the opposite side of the divide. Yet as the road winds it is 175.

And so the distances deceive. For a man afoot in the mountains, such deception is particularly frustrating. A hiker on level ground can cover perhaps 20 miles a day comfortably. In the mountains he may be lucky to make five. The very clarity of the air confounds him, so that a far summit glimpsed from the trailhead appears to be a 30-minute walk-up. As the hiker climbs toward it, the crest disappears behind intervening ridges. An hour or two later and several gullies farther on, it may show up again—only to seem as far away as it did at the start. By late afternoon the hiker might arrive, his muscles shaky, his head light from the altitude, a cramp where his backpack weighs against his shoulder, his exhausted legs behaving in strange and unpredictable ways.

More unpleasant moments have been spent in the mountains by people underestimating distances than for any other reason. Captain Zebulon Pike is a classic example. Pike evidently had less sense of both distance and direction than any other famous explorer in the history of the West. He set out for the Rockies in 1806, following the Arkansas River upstream across the Great Plains. On November 15 he sighted his first mountain, shimmering on the western horizon like "a small blue cloud." A week later and 100 miles farther on, he had neared the foothills. He estimated that a day's march would take him to the mountain's top. He was wrong. The elusive peak rose 50 miles ahead.

He pressed on with three companions, wearing a light summer uniform and, for some reason, no socks. Three days later he was still floundering around the foothills, in snow up to his waist, without food, blankets or water. On Thanksgiving Day, November 27, 1806, he reached his highest point, the crest of a foothill that gave him a superb view of the summit, 12 miles away. But a labyrinth of steep ravines blocked the way. One look told him it was time to quit. It was "as high again as what we had ascended," he scribbled in his journal, "and it would have taken a whole day's march to have arrived at its base, when I believe no human being could have ascended to its pinical *(sic)*."

Wrong again. Pikes Peak was climbed in 1820 by three members of a surveying expedition led by Major Stephen Long, who calculated its height as 11,507.5 feet. Long was a bit conservative; the mountain's true height is 14,110 feet. But he came closer than Pike, who thought the summit to be 18,581 feet high.

Pikes Peak has been climbed since then with ever-increasing regularity. The first woman to reach its top was Julia Archibald Holmes of Lawrence, Kansas, 20 years old and newly married, who hiked up it for a lark in 1858. She wore a calico skirt and bloomers, packed a copy of Emerson's *Essays* and a writing kit, and dragged her husband along too. On the summit, crouching behind a boulder from a chilling wind, she penned letters to a number of close friends. Just being there, she wrote, "fills the mind with infinitude, and sends the soul to God."

As times grew less pious and more macadamized a road was built up the mountain, and today most thoughts of infinitude on Pikes Peak relate to the size of the crowd. Sightseers throng up it at the rate of 350,000 a year; an annual sports car rally goes to the top, and even a marathon run. Record time for the run up—13 miles—is 2 hours, 9 minutes and 30 seconds, set in 1972 by a 21-year-old Californian. The race also has a Superman Division for runners aged 70 and above.

In the wild and unpaved West of 1806, it was a miracle that Pike and his fellow climbers neither starved nor froze to death. Then, as now, Rocky Mountain winters could be horrendous. At Rogers Pass in Montana, 140 feet below the summit of the divide and half a mile to the east, the temperature on January 20, 1954, dropped to 69.7° below zero. It was the coldest moment, south of Alaska, in U.S. Weather Bureau history. Snow, too, can pile to mighty depths. Seventy-six inches fell near the divide above Boulder, Colorado, in a single 24-hour period during a blizzard that began on April 14, 1921. Another national record.

In 1871 a hardy English lady named Isabella Bird blundered into a typical Rocky Mountain blizzard. She was 42 years old, a spinster, and she was touring Colorado on horseback, clad in the same light riding dress she had worn the previous summer in Hawaii. One December morning she headed down toward the plains from Estes Park, in Colorado's Front Range, when the storm hit. "I faced a furious east wind," she wrote later, "loaded with fine, hard-frozen crystals, which literally made my face bleed. . . . It grew worse and worse. I had wrapped up my face, but the sharp hard crystals now beat on my eyes—the only exposed part—bringing tears into them, which froze and closed up my eyelids at once. I had to take off one glove to pick one eye open, for as to the other, the storm beat so savagely against it that I left it frozen, and drew over it the double piece of flannel which protected my face. I could hardly keep the other open by picking the ice from it constantly with my numb fingers. It was truly awful at the time."

Yet winter, for some people, is the most beautiful time in the mountains. Between blizzards, when the sky clears and the wind drops, and the mountain light gives a fine blue transparency to the shadows between the drifts, the landscape seems wonderfully benign. Even in freezing temperatures you can walk about in shirt sleeves and not feel cold. The air at high altitudes is so thin and dry that it filters out comparatively few of the sun's rays, which pour so generously over your skin that the real danger is not cold, but sunburn. The rarefied air accounts for another mountain phenomenon: the intensity of visible light. It floods down in concentrations 25 per cent stronger than light at sea level. Thus the illusory closeness of faraway land masses; no haze of atmosphere sets them into perspective. Thus too the clarity of each pine needle and snow crystal, and the extraordinary brilliance of the sky.

The snow itself brings order to the chaos of landscape. It softens the contours, building along the lee side of ridges in luxuriously curved overhangs, called cornices, that look as though an ocean wave had fro-

Storm clouds glower over the Great Divide in a part of northern New Mexico where the watershed is defined merely by a low hilltop.

zen in place just before breaking. It obliterates ditches and tree stumps and boulder fields, so that a man on skis or snowshoes can skim over obstacles that would make for tough hiking in summer. Snow looks right and feels right. It is the persistent reality of the high country, the source material for the rivers that emanate from the Great Divide.

Snow brought me my own discovery of the Great Divide. I had come to Vail, Colorado, during an April vacation to ski. The season was ending, not for lack of something to ski on (the resort where I stayed had reported a snow base 93 inches deep), but for lack of customers. So much snow had fallen for so many months that people were tired of the stuff. I found this attitude incomprehensible, like a miser who sees money turned down. Snow in Vermont, where I usually ski, comes and goes according to its own strange whims, and never in the powdery extravagance I saw here. I swooped down the trails like a boy with a Christmas sled, past aspen groves and stands of lodgepole pine, down broad open bowls where the snow seemed as light as sea foam. Perspectives were vast. Skiers on the opposite side of the bowls looked like tiny mosquitoes on a white wall.

But a far grander presence tugged at my attention. My eyes kept lifting above the tree line, past the gondola towers and warming huts, to a saw-toothed ridge in the far, high distance. It stood aloof to the point of arrogance, well above the concerns of skiers, resort owners and mankind in general. It was part of a range, I was told, that abutted the Continental Divide. Could one get there? Yes, indeed. And with a bit of effort, one could walk along the divide itself. The journey would require a day's hike on cross-country skis, an overnight camp-out near the timberline and a climb to the heights the following day.

I am the kind of person who prepares for strenuous exercise by reading a book. Thus, in anticipation of my hike, I settled into an armchair by the lodge fireplace to learn some facts about the divide. It is essential, I discovered, to distinguish between the Continental Divide—a geographer's line separating drainage basins—and the mountains it runs through. The Rockies are part of a vast system of highlands called the Western cordillera that runs down the western sides of both North and South America. The Rockies themselves include at least 60 smaller ranges, sprawling from northern Canada through New Mexico.

These mountain systems are intermittent; in places the general upthrust gives way to grassy parkland, prairie and desert. The divide itself weaves through them like a Saturday night drunk, twisting, turning, backtracking from one mountain ridge to the next. Its northern ter-

Wrapped in icy shrouds fashioned by a sudden freezing rain, three withered wild flowers brave winter's onset in a high Rockies meadow. Their roots, protected by a blanket of snow, will survive the winter, and the flowers will bloom again when summer comes.

minus is above the Arctic Circle in Alaska's Brooks Range. So for the first 4,000 miles or so, I was surprised to discover, the divide separates the Pacific drainages from the rivers that empty north into the Arctic Ocean. Then, at a point in Jasper National Park in the Canadian Rockies, the Arctic drainage ends. From here south, all rivers flow toward either the Pacific or the Atlantic. The divide rolls down through Canada, across the United States and Mexico, dips toward sea level at the Isthmus of Panama and rises along the spine of the Andes to the tip of South America. No one knows exactly how long it is, for parts of it, in the Andes, are still blanks on the map. But a fair estimate, counting all its perambulations, might be 25,000 miles, or slightly more than the distance around the globe. Even within the United States, the length of the divide is formidable, approximately 3,000 miles from the Canadian border to Mexico. I would have to limit myself to a visit here in Colorado, and a few selected spots in Wyoming and Montana.

I turned from geography to history. The divide marked the western extremity of the Louisiana Purchase, I learned, and thus in 1803 became a boundary of the United States. Two years later Americans crossed it for the first time when two Army officers, Meriwether Lewis and William Clark, were sent by President Thomas Jefferson to see what the Western lands were all about. For the next five decades nobody ventured into the mountains much except explorers and beaver trappers. Just about everyone else crossed over as quickly as they could to Oregon and California, where the living was easier and a man might get rich in the California gold fields.

Then, in 1858, a prospector back from the California gold rush raised color in his pail at a creek near present-day Denver, Colorado. A new gold rush: Pikes Peak or Bust. (Geography was nobody's long suit in those days; Pikes Peak is a full 70 miles south of Denver.) For the next 40 years, anyone who could get his hands on a shovel and a pack mule, it seemed, tromped through the Rockies looking for pay dirt. Log-cabin mining camps sprang up along the length of the divide: Alder Gulch, Virginia City in Montana; Atlantic City in Wyoming; Cripple Creek, Aspen, Central City in Colorado. If you missed the gold, silver would do, or copper, or lead, or you could bring in cattle to feed the camps.

But ores play out, and metal prices drop, and by the turn of the century most of the camps had bcen abandoned to the wind. There were exceptions, of course. Zinc, lead and molybdenum are still blasted from the earth near the resort where I was skiing. And about three miles west of the divide in Butte, Montana, the world's most famous copper mine has poured some two or three billion dollars into the bank account of the Anaconda Company, established in 1874. But for the most part, by 1900 the land along the divide had reverted to wilderness.

A few days after absorbing these facts, I set out for the divide on a brilliant Rocky Mountain morning in the company of a guide. Snug against my shoulders I carried a backpack stuffed with about 40 pounds of sweaters, cameras, suntan lotion, cooking pots, sleeping bag and part of our tent. Strapped to my feet were a pair of cross-country skis that seemed far too slender and delicate to support so much weight. My guide, Bruce Batting of Vail, Colorado, hefted an even bulkier pack. He is a powerful man, but short, and his pack seemed nearly as big as he was. When loaded up he looked, from the back, like a walking siege tower. Bruce does not talk much. His remarks, when he makes them, tend to be practical and specific. He says things like "Tracks, bobcat" or "Careful there, avalanche slope" and then keeps on walking.

South of Vail the Continental Divide makes one of its innumerable shifts of direction. It twists in from the east, rising through a mountain group known as Chicago Ridge, and then dips below the timberline to 10,450 feet at Tennessee Pass, where it is breached by the main north-south highway from Vail to the mining town of Leadville. Just west of the highway it climbs again, until it reaches a squat stone pyramid 13,211 feet high called Homestake Peak. Here it turns abruptly south, to thread through the highest mountain range in all the Rockies, the Sawatch. Our plan was to drive to Tennessee Pass, leave the car, and strike out cross country parallel to the divide for Homestake Peak. A day's hike would bring us to the base of Homestake, where we would camp for the night. The following morning we would assault the peak, then ski down to the car by late afternoon.

The road to Tennessee Pass snakes upward across the mountain faces, where exposed cliffs reveal rock strata in variegated colors; alternating layers of maroon, coffee, apricot and buff attest to the intermingling of different minerals. As the road gains altitude it also flattens, crossing through upland meadows and quietly undulating hills. The pass itself is simply a low saddle between two hills—not at all the spectacular cut that most people expect of a breach in a mountain chain, and surely not on the Great Divide. But this is the way with passes; they usually occur where the terrain is most easily crossed.

I found a gentle charm, however, in Tennessee Pass. The hills on either side slope upward invitingly, with snow-covered glades penetrating the generous clumps of evergreens. Most of the trees were lodgepole pines. Their tall thin trunks rose straight as ship's masts, and their olive-green branches curved upward at the ends. At each tip a cluster of khaki-colored cones surrounded a tight bundle of reddish knobs. These were the spring growth of lodgepole flowers (dendrologists call them strobiles), the male elements in the odd reproductive cycle that would pollinate the female cones. The cones, though fertilized, remain tightly closed on the tree until extreme heat, usually a forest fire, causes them to open up and drop their seeds.

We climbed the hill west of the pass, skiing easily through the lodgepoles. No trail led where we would go, but that did not matter; the snow would pave over most obstacles. Its surface was dry and fluffy —there had been a storm the day before—and neatly grouped tracks of snowshoe rabbits crossed it every which way. The rabbits themselves are hard to see. Each autumn they shed their brown summer coats and grow winter disguises of white fur. The camouflage is not perfect, for the fur on a snowshoe's feet is out of sync with the rest of him. It changes six months late, so in summer he has white feet and in winter, brown.

Climbing steadily, we followed the divide through open glades where the sun reflected from the snow with such dazzling intensity that my eyes ached. Occasionally a gray tree stump poked like an iron chimney through the surface. Late in the 19th Century, when prospectors mined gold along this stretch of the divide, they cut timber to build sluices and to shore up tunnels. These early prospectors protected their eyes by painting their cheekbones with a mixture of bear grease and wood ash, and by squinting. A chancy method, it seemed to me. I became convinced that without my goggles I would have gone snow blind. "Don't worry," Bruce told me, "it only lasts a couple of days."

In the shade of the forest the hazards were different. Lodgepoles sometimes have a tendency to crowd together so that the branches of each tree tangle with those of its neighbors. It became impossible to ski 20 yards without getting deeply entrapped. Adding to our difficulties was the unpredictable character of the snow, which had been firm and uniform in the open fields. But here in the forest the snow piled in irregular mounds, the sun coming through to melt it in patches.

In some spots the snow was packed hard; in others it had turned to oatmeal mush. To compound the problem, temperatures had risen enough during the past week so that on the warmer, south-facing slopes a premature spring thaw had begun. Snow, though most people do not realize it, evaporates faster than it melts. It does so in such a way as to leave air cavities under the surface, like holes in a sponge. When the weather is cold, this sponginess is no problem; the firm snow that remains is usually strong enough to support a man's weight. But when the temperature rises, the structure weakens and the snow turns rotten. Ski across a pocket of rotten snow and the whole thing gives way, collapsing you into an impossible tangle of twisted limbs and unmanageable gear. The only way out is to remove your pack, unbuckle your skis and flounder up the edge of the pit.

The hike through the forest lasted most of the day. At one point we emerged onto a man-made ditch that cut straight across the divide. I asked Bruce what it was for. "Takes water from one side of the divide and pipes it to the other," he said, and lapsed into silence. We skied on through the forest and into the afternoon, the rays of the lowering sun glinting and diffusing among the pine needles. Suddenly the trees opened out, unveiling in the distance something icy blue, pristine and immense: the high ridges of the divide, more than 13,000 feet above sea level, where we would stand the following day.

We left the divide in search of a stream where we could camp. We found the north fork of West Tennessee Creek flowing along an easy bottomland where the lodgepoles thinned out and a new kind of tree took over. It was Engelmann spruce, a majestic evergreen that survives in a narrow band of about 1,000 feet just below timberline. Its trunk is a long straight cylinder, rich orange in color, and its branches droop in heavy emerald garlands to the snow. Some stood well over 100 feet tall with trunks almost a yard in diameter. These were old trees. In the short growing season at high altitudes, it had taken them 400 to 500 years to reach these stately dimensions. It seemed a fitting life span for such imperial beings.

The spruces along the creek stood at nicely spaced intervals, like trees in an arboretum. We set our tent between two of them, on a patch of evergreen needles where the snow had drifted clear. Indeed, we found cozy snow-free pockets at the bases of most of the spruces, whose lower branches swept the ground, forming a kind of natural tent to keep out the elements. "Good place to go if you're caught in a blizzard," Bruce said. "Animals do it. Bears hibernate there, and you can sometimes find places where elk and mule deer have bedded down." He had become positively loquacious.

There is a natural impulse, when you shuffle off your backpack, to let the muscles go slack, ease the ligaments and allow the blood to run freely into all the capillaries it may have missed while you were hiking. The mind itself relaxes, with thoughts of fire, supper and a warm sleeping bag. Nothing could be less appropriate. For a tent must be pitched, a fire pit dug, sleeping bags must be unrolled and water fetched. This last was my job. I trudged through the snow on my skis to the creek, about 200 yards away.

The creek ran in and out of snow cover, like a road through underpasses. It would emerge from a snowbank, swirl about in a little pool and disappear under the arch of the next snowbank. I removed my skis and crawled on my belly to the edge of a pool, trying to avoid an overhang that might collapse and dump me into the water. It felt good to lie flat for a change, and my gaze flowed out with the water as it dashed among the rocks in the pool and swept under a snow bridge. A dark shape flickered in the water: a trout. I could just make out its speckled back before it swam into the shadow of an overhang.

I dipped my hand into the water and retracted it immediately—a reflex against the cold. The creek was barely above freezing. I then lowered the canteen into the current, holding it gingerly with my fingertips. Too gingerly. The current, surprisingly swift, wrenched away the canteen and swept it downstream under the snowbank. There was no retrieving it. Had Bruce not been carrying a second canteen, we would have spent the rest of the trip without water.

Winter along the Great Divide allows few such mistakes. Men have frozen, gone insane, starved to death, and on several occasions they have carved up their traveling companions and roasted them for dinner. One mountain man who used this latter technique was a prospector named Alfred Packer. In early winter of 1874, Packer set out with five other men to locate a mining claim in the San Juan Mountains in southern Colorado. On the flank of the divide they were caught in a blizzard. All hands were counted as lost until the following spring when Packer showed up, sleek and sassy but minus his friends. The circumstances were sufficiently incriminating—the party's winter camp was strewn with well-chewed human bones—that Packer was brought to trial on charges of "eating up the Democratic majority in Hinsdale county." Packer served 17 years of a 40-year sentence, but was then let out on the grounds that he had committed not murder but only cannibalism.

Zebulon Pike, under different circumstances, left a small legacy of bones to the Rocky Mountain winter. After failing to climb his mountain, and mistaking the Arkansas River for another waterway several hundred miles away, he spent the next few weeks of the winter of 1806-1807 slogging through the fastnesses of southern Colorado, his entire company still clad in cotton summer uniforms. One by one his soldiers froze their feet, and he had to leave some of them bivouacked in the mountains. Later he sent rescue parties to carry them out. An entry in his journal describes the situation. "This evening the corporal and three of the men arrived," he wrote. "They informed me that two men would arrive the next day . . . but that the other two . . . were unable to come. . . . They sent on to me some of the bones taken out of their feet, and conjured me by all that was sacred, not to leave them to perish far from the civilized world. Ah! Little did they know my heart, if they could suspect me of conduct so ungenerous."

A man-made snowslide roars down Avalanche Corridor Irene near Silverton, Colorado. Forest Service snow rangers fire artillery shells at slopes like this to trigger harmless, though spectacular, slides before enough snow can pile up to precipitate a disastrous natural avalanche.

Even today people freeze to death along the divide. Each winter a few motorists neglect the warnings of the state highway departments and perish in blizzards in the high passes. But the most persistent danger is something entirely different: the avalanche. It seems hard to believe that the beautiful white fluff that covers the mountains in such soft and generous quantities may bring death. But it can and does. On certain mountain faces, where the slope is sufficient, the wind and temperature just right, the snow cover can give way, quite suddenly and without warning. Thousands of tons of white destruction hurtle down the mountain face at express-train speeds. Sometimes the entire slope is swept bare—trees, boulders, houses, skiers, everything, carried down to the valley and buried until the following spring. Slides are so frequent at Loveland Pass in Colorado where the old highway from Denver crosses the divide that the major avalanche corridors have been given names: Black Widow, Grizzly, Little Professor, Associate Little Professor, Five Car and Happy End.

If caught on skis in an avalanche, you can try to ride it out, like a surfer riding a wave. If knocked down you can sometimes survive by swimming to the top. (A Husky dog, with whom I became friends at Vail, was once caught in a slide, buried for three days and dug himself out, happy but a bit hungry.) But the chances are you will be swamped, suffocated or ground to bits by chunks of compacted snow. Here is what happened on February 28, 1902, when a trinity of avalanches tore apart a mine at Telluride in the San Juans. Historian David Lavender, a native of Telluride, describes the aftermath of the first slide: "As rescuers toiled up from town, a second avalanche dropped on them. And finally, after the rescue work had ended and the weary volunteers were

dragging the corpses and injured back to town on sleds, a third slide fired a parting shot. The triple blows killed 19 men, including my grandfather's brother, and cruelly hurt 11 more. My stepfather-to-be saved himself by seizing a tree as he was being hurled down a hillside; then he clung, deafened and semiconscious, almost suffocated by the snow that had packed like cement into his ears, mouth, and nose."

Yet now at our camp, as we sat cross-legged by a warm and fragrant fire of spruce boughs, the light of an ascending moon bringing out the contours of Homestake Peak, the mountains seemed benign and beautiful. As the night deepened, the sky took on an indigo clarity I had never seen before. It seemed unimaginably far, yet close enough to climb to. The moon, a sharp crescent, was so bright its light suffused an entire quadrant of the sky, obliterating the lesser stars, and virtually extinguishing our fire by comparison. The snow itself seemed to glow with an internal incandescence. Later, when I poked my head out of the tent into the cold darkness, the moon had set. The stars, without competition, flashed as thick as confetti. I noticed something I had read about in textbooks but had never seen: heavenly bodies have color. Mars is indeed red and Sirius, the Dog Star, distinctly blue.

We breakfasted soon after sunrise on oatmeal, hot chocolate and ham. Bruce packed a rucksack with a full canteen, chocolate bars, first-aid kit, wax for skis, ice ax and a coil of bright red string called avalanche cord. The theory behind avalanche cord is that you tie one end to yourself if you have to cross an avalanche slope. Then if the snow breaks loose and buries you, one end will protrude, allowing your companion to find you and dig you out.

Our camp was close to timberline, at about 11,000 feet. To reach the top of Homestake we would have to gain 2,211 feet of elevation. We glided up through the Engelmann spruces along the creek bottom and then began to climb the shoulder of the mountain. The trees grew steadily thicker. An occasional gnarled and weathered stump protruded from the snow like a monstrous, prehistoric head. We became aware of an intermittent murmur in the treetops. It grew deeper and stronger as we ascended until it became a roar, as powerful as an engine: the wind funneling across the roof of the divide. Through the farthest tree trunks we could see the white of open space. The trees thinned again, became stunted, stopped. This was the timberline. Nothing ahead but the open snow, the wind and the smooth white pyramid of the peak. We zipped up our parkas and set out.

There was life under the snow: dormant grasses, sedges, lichens, the roots of tiny alpine flowers thrusting into the rock, nests of hibernating marmots—a special, delicate world of high-altitude tundra that exists only above the line where ordinary trees can no longer grow. But we could see little evidence of it. Here and there the miniature bough of a stunted demi-tree elbowed above the snow like a mummified arm. These were dwarf versions of the same giant spruces growing below us. Here above the true timberline they manage to survive by hugging the ground, their foliage protected by snow cover from the killing desiccation of the winter wind. But this was all.

We had to lean hard into the wind to keep our feet. It swept down from the peaks at an erratic 40 miles an hour, stopping, starting, accelerating, shifting gears as it skidded across the convolutions of the landscape. It lifted a banner of white powder from a high ridge and sculpted the snow into a cornice on the lee side. The surface we skied across had also been carved by the wind. The snow was packed hard, and it rippled like the corrugations in a tin roof. Windblown crystals stung our eyes and cheeks, and invisible hands tugged our pants legs as though trying to pull us over backward.

The climb would take several hours. An interminable series of switchbacks would carry us upward yard by yard for 1,500 feet. If you closed your eyes to the sun, your ears to the wind, and sealed your mind against the beauty of the surrounding peaks, you might have imagined yourself walking sideways up a staircase, wearing swim fins, to the top of a building 150 stories high. But you could not do this. Despite fatigue, despite the glare from the sun and the dizziness from oxygen lack, despite the throb in your temples with each step, you could not shut down your senses. With each yard gained the mountain landscape opened out. Above us the ridges of the divide ran off each flank of Homestake. Bare rock broke through the snow, sharp and intricate as the fenestrations of a Gothic church. Between the outcrops snow lapped down in soft lyric arcs, gliding and dipping in a rhythm that seemed almost musical, like the phrasing of a cello. Wind had cleaned the snow from the tip of Homestake, revealing a crown of heaped, angular boulders at its summit. Here, 13,211 feet above the ocean tides, the great barrier reached its apex.

Mountains Assaulted by Ice

PHOTOGRAPHS BY DAN BUDNIK

Viewed from far off, the cloud-wrapped Rockies in the Great Divide country of Colorado and Wyoming appear serene and immutable. Seen close up, however, the mountains betray the scars of an age-old battle that still rages, slowly but steadily changing their aspect. It is a battle between rock and ice.

In these hostilities a major weapon is the unceasing barrage of snow to which the Rockies are subjected. Ever since they were pushed up to their present eminence by upheavals that occurred about five million years ago, the mountains have stood as a barricade against the prevailing westerly winds. Confronted with this rampart, the moving air rises abruptly and cools, losing its ability to hold water vapor. Most of the resulting precipitation is snow, which amounts to about 200 feet a year.

Throughout much of their existence the mountains have endured a series of ice ages. During these extended periods of cold, when the year-round temperatures averaged well below freezing, the abundant snow compacted into ice in sheltered places near the mountaintops. As the snow continued to pile up, the ice masses grew and, in the form of glaciers, began to move down the mountains' flanks to fill up the lowlands at their feet. Every time these incredibly ponderous ice masses ground their way down they left their marks, from their origins high in the mountains to the many ravines and valleys that they gouged and widened as they found the easiest routes down the slopes.

Each period of glaciation has been followed by a warming trend during which the glaciers have receded back up toward their mountaintop incubators. Such a trend prevails today, but even so there are still scores of glaciers in the Rockies. They are relatively small and of recent vintage, having been formed during a short cold spell only 300 years ago.

There have been several of these mini-ice ages during the span of recorded history, but they have been merely brief lapses in the comparatively mild climate of the modern era. Indeed, most of today's young glaciers are shrinking because of the warm summers; and yet the eight-month-long winters are still cold enough so that the conflict between rock and ice continues. Less dramatically now, but inexorably, the ice keeps chipping away, eroding and weathering the gneiss and granite of the Rockies and altering their look.

Peaks of Colorado's San Juan range pierce a fleecy sea of stratocumulus clouds. During the most recent ice age, 12,000 years ago, glaciers enveloped these mountains to approximately the same height as this cloud level.

A glacial trough, at left, illustrates the effect of a glacier's abrasive passage.

Bowls and Troughs Gouged by Glaciers

From an airplane over the Rockies the most dramatic and recognizable signs of former glacial activity are the immense bowls, high in the peaks, from which the rivers of ice began their downward flow. Called cirques, these bowls start as shallow pockets, usually on the lee side of the topmost ridges, where snow collects and is protected from the winds whistling overhead.

If temperatures are cold enough, the snow lingers and eventually compacts into ice, molded to the rock it lies on. Further accumulations of snow enlarge the ice mass until it is hundreds of feet thick. Its ponderous weight exerts intense pressure on the lowest layers of ice, so that they gain a firm and lasting hold on the rock. Then, as gravity starts it moving, the glacier plucks large pieces out of the floor and walls of the cirque, and carries them away in its icy grip. Thus, bit by bit, the glacier excavates the cirque, gradually making its back wall higher and steeper.

When the glacier outgrows its cradle, it spills over the cirque's lower lip and starts slowly moving down the slope, at a rate of perhaps a few hundred feet a year. Following the path of least resistance—usually a river valley or a broad canyon—the glacier abrades as it goes, the boulders imprisoned in it scouring and rounding the valley. The result is a widened trough *(left)* that remains as the glacier's distinctive signature.

The rounded backs of glacial bowls, or cirques, dominate this aerial panorama of a section of Colorado's Rocky Mountain National Park.

Originally excavated by the erosive effect of glaciers, this mile-wide cirque shows evidence of frost action in the jaggedness of its walls.

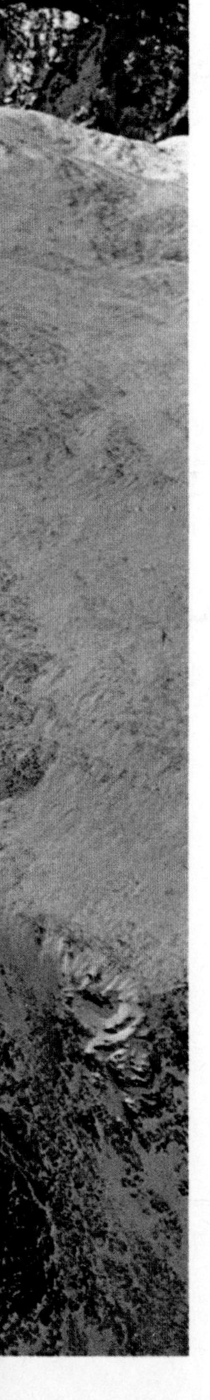

The Continuing Attack by Frost Action

Though the glaciers have gone from most of the cirques they dug out of the mountains, the interiors of the bowls remain under steady assault by ice, in the form of frost. Rain and melted snow collect in crevices, then freeze and expand to shatter rocks, widen fissures and deepen notches. The softer strata of rock are most vulnerable, of course, but even the most obdurate granite and quartzite finally succumb.

This relentless weathering upon cirques continues the enlarging process begun by the glaciers. A cirque flanking a relatively broad ridge, like the one shown at left, simply gets bigger around as frost action chips away at the walls. But a more arresting pattern emerges where several cirques were originally carved out back to back, as seen in the photograph at right. The backs of two of the cirques form a common steep-walled ridge, center, called an arete. As frost action continues through the centuries, breaking down the top of the arete piece by piece, the ridge will become ever more saw-toothed, and in time it will disintegrate altogether, leaving in its place a pass, known to mountaineers as a col.

In the case of the arete shown here, the result of that slow, inevitable process will be of more than casual interest. The Great Divide now runs along the top of the ridge, and the crumbling of the arete will alter, however slightly, the course of the continental watershed.

Four back-to-back cirques in the San Juans define a narrow ridge, or arete.

Cirque lakes (foreground and right) and paternoster lakes (top left), all reminders of past glaciers, nestle on either side of the Great Divide

in Wyoming's Wind River Range. The divide snakes along the winding ridge at center.

A Legacy of Lakes High in the Peaks

Scattered throughout the highest reaches of the Rockies like so many dark gems are countless mountain lakes whose beds were dug by past glacial action. Some of them, like the half-frozen pond in the foreground of this picture, lie isolated in bowls once occupied by glaciers and are called cirque lakes, or tarns.

Others lie in tandem along the line of a former glacier's slow march. Each marks a place where the ice ran into a resistant patch of rock and was forced to move horizontally for a bit before heading downhill again. Thus was formed a series of shallow depressions—like worn treads in a stone staircase—which have filled with water and become interconnected. The curious name by which they are known—paternoster lakes—stems from the fact that they are strung along a single watery thread like beads on a rosary.

The water in these lakes comes mainly from snow, as did the glaciers' ice; snowfall is probably as plentiful today as during the last ice age. But now with temperatures warmer, much of the snow melts in summer instead of remaining year round, and runs off through the paternoster lakes, as each overflows into the one below. But when the Great Divide separates two lakes, even two as close together as those pictured here, their waters cannot mingle; instead they head off from their common mountain for opposite shores of the continent.

A Carpet of Trees Where Glaciers Grew

In the warming climate of modern times, the snow line (the lowest limit of year-round snow) has retreated back up the Rockies' slopes. In its wake trees have followed to encroach on what was once glacial territory, gradually reclothing the Rockies in green instead of white.

Still, only the most rugged species of trees, such as alpine firs and Engelmann spruces, can grow at these altitudes—in some locations over 11,000 feet. At that, they are often stunted and bent, deformed by the harshness of their environment.

Even as they take over the glaciers' former domain, the conifers owe them a debt, because the trees find their best protection from cold and wind in areas such as glacial troughs and cirques carved out by the ice. Paradoxically, lower on the mountains where the trees are generally well established, some slopes have been made uninhabitable as a result of frost action. The bare brownish hillside at the center of this picture is a talus slope, consisting of masses of ice-chipped rock fragments. Constantly sliding in a slow movement called talus creep, the slope is too unstable for trees to get a foothold.

Where they are able to dig in, the trees send their roots into cracks and fracture the rock, breaking it into smaller pieces. Thus in their own way they contribute, like the ice, to the never-ending process of wearing down the mountains.

Reaching up toward the snowy crest of a mountain in the San Juan range, fir trees bring

life to a slope once shrouded in glacial ice. The trees grow thickly in ravines the glaciers carved, but thin out along unprotected ridges.

2/ Of Rocks and Rivers

The fall of dropping water wears away the stone.

LUCRETIUS/ *DE RERUM NATURA*

A circular bronze plate is set into the rock at the summit of Homestake Peak, on the Continental Divide in the Colorado Rockies. It was bolted into place some years ago by a U.S. Coast and Geodetic Survey team and was later used as a reference point while making a map. The surveyors and cartographers must have had quite a job. For the aspect of the land from Homestake Peak is utter, primordial chaos. The mountains heave up with the erratic violence of a storm at sea, the winter wind lashing a spume of snow from each shoulder and crest. Ridges merge and separate without apparent reason, outcrops shoot up helter-skelter, valleys cut between the peaks in sudden bewildering gashes. Who could ever make sense of it?

Someone, certainly, for a map does exist. And here at the top of Homestake Peak, if you can hold your copy of it steady against the tearing of the wind, the map will show you some rather interesting things. It will help you sort out the turbulence of landscape into separate mountain ranges. It will establish the position of the rolling parklike valleys that lie between them. It will show you the paths of rivers as they course down either side of the Great Divide. And to an experienced eye, it can be an important tool in deciphering the geologic past of the mountains themselves.

The most unmistakable features are the high ones. From where you stand on Homestake, you can count the number of tall peaks that break above 14,000 feet. You can see a total of 22, or almost a third of the 14,000-foot mountains in the continental United States south of Alaska. For Colorado is high country, as natives are generally quick to point out, and it contains the loftiest summits in the Rocky Mountain system. All of the chain's 53 Fourteeners rise here, and the mountains that crest above 10,000 feet number 1,140, give or take a few dozen. Homestake Peak is part of a range called the Sawatch, which carries the Great Divide southward from where you stand. It includes the highest crag of all, 14,433-foot Mt. Elbert, and scores of others, sailing into the distance like a row of battleships.

Northwest of Homestake Peak the land drops off steeply into a narrow valley, and then rises to another cluster of peaks on the far side. One of these summits, an ungainly granite protuberance 14,005 feet high called the Mount of the Holy Cross, has a certain degree of fame attached to it. One face has weathered in such a way that snow lingers in its crevasses in the form of a giant cross. In an age of greater piety than ours this phenomenon was deemed significant. Various miraculous rescues were attributed to it, artists painted it, Henry Wadsworth Longfellow wrote a poem about it, and for several years during the 1930s an evangelical preacher conducted faith-healing services from a vantage point on the mountain next to it.

You cannot see the snowy cross from Homestake because the cluster of intervening mountains hides it. What you do see is to me just as fascinating, for it offers a particularly valuable clue in sorting out the landscape. If you bring your eye back down from the mountains toward the intervening valley you will notice, among the spruce cover, a patch of open snow. By consulting your map, you discover that this open space is actually a frozen lake. A trickle of water feeds from the lake into Homestake Creek, hidden among the spruces at the valley bottom. From there it flows into the Eagle River, a tributary of the Colorado, and so to the Gulf of California. Thus from the top of Homestake Peak you can make out, as surely as water flows downhill, the Pacific slope of the Great Divide.

If you face east, the story repeats itself. From a rock amphitheater on the east flank of Homestake, another lake nourishes Tennessee Creek. Several miles downstream, in a hazy brown valley called Tennessee Park, the creek mingles with other mountain rivulets to form the headwaters of the Arkansas River. You can just make out the glint of sunlight on the river's surface. And if your map is extensive enough, you can trace the route of the Arkansas as it glides southward through its own valley, past mine shafts, ghost towns, cattle ranches and irrigated farmlands. At the foot of the valley, having grown into a fairly rambunctious torrent, the river cuts eastward through one of the most spectacular canyons on earth, the Royal Gorge of the Arkansas. It then gushes out onto the Great Plains, and eventually joins the Mississippi on its way to the Gulf of Mexico.

The rivers are there because of the mountains, which in a sense provide the substance that fills them. As you stand among the high peaks on this winter day, you can see this raw material of rivers all about you, filling the valleys and blanketing the lakes and whitening the rocks of the highest crags. It is snow, which accumulates here in vast quantities and supplies 75 per cent of the water in Colorado's mountain streams. The prevailing winds come in from the west. When the winds hit the mountain slopes they rise and drop their moisture among the peaks and along the Great Divide ridges. Thus the mountains constitute a weather barrier that captures a good part of the snowfall in the American West. Because these mountains are high and cold, they keep their snow throughout a good part of the year. They act like a huge frozen reservoir that releases its water bit by bit as the snow gradually melts during spring and summer, and thus helps keep the rivers running. And as with most big reservoirs, the snow in the divide watersheds has considerable economic value.

Most people when asked to name the natural treasures in the Rocky Mountains will start talking about minerals—gold, silver, gem stones and so on. They are only half right; for snow, which they generally forget to mention, is the most precious material of all. Back in the 18th Century, before anyone had heard of the Rocky Mountains, a Mississippi River explorer named Jonathan Carver was told by the Indians of a miraculous range to the west "called the Shining Mountains from an infinite number of crystal stones of amazing size, with which they are covered." Given the congenital optimism of the day, Carver assumed that the crystals were diamonds, and that the Shining Mountains would yield vast riches to whoever found them. They have indeed, for Jonathan Carver's crystals were of course the snow along the divide watersheds. Who uses it, and how, is a matter of grim contention. The battle rages today in state legislatures, law courts, the U.S. Forest Service and the Department of the Interior.

Most of the American West is dry. There is just not enough water for everyone. Only 14 inches of precipitation fall in Denver, Colorado, in a normal year. This is not nearly sufficient to fill the dishwashers, flush the toilets and sprinkle the front lawns of the city's million or so residents, much less irrigate the wheat, sugar beets and potatoes in the surrounding farmlands. The same holds true in Colorado Springs and Pueblo to the south, and in most communities along the western edge of the Colorado Plains. Some of their water comes from local wells; but most of it must be drawn from rivers and lakes in the mountains. When the rivers and lakes run low, the plainsmen look for other ways to tap the watersheds along the divide.

One solution—and the source of all the controversy—is to import water from the wetter country west of the divide. Thus copious sums of money have been spent tunneling through the mountains, under the divide, to pipe out water from lakes and reservoirs on the west slope and bring it east to the plains. Denver takes 792 million gallons a month from Dillon Reservoir on the west slope and pumps it under the divide via a tunnel blasted 23 miles through mountain bedrock. Two other tunnels add to the flow. But the city grows thirstier each year, and the Denver Water Board keeps urging yet another transmountain viaduct.

A slew of different communities claim the runoff from Homestake Peak. Ever since 1932 the town of Pueblo has been using water from a ditch that cuts through a flat section of the divide near Tennessee Pass. This was Bruce Batting's ditch; I had been forced to cross it to reach Homestake, removing my skis and picking my way over a rickety wooden plank that served as a bridge. More recently a reservoir was constructed in the west flank of Homestake, and its water was piped under the Continental Divide to Colorado Springs and Aurora, a Denver suburb. Then in 1972 another diversion project was completed, for $300 million, that drains water from two rivers on the west slope of the Sawatch Range. The diversion was not entirely popular. The towns of Aspen and Glenwood Springs lie along these two west-slope rivers, and like to use the water themselves.

People west of the divide regard these diversion projects as an outrageous aquatic burglary, and speak of them with venomous hatred. Such tinkering with natural watersheds, they claim, will wreck the delicate ecological balance along the divide. Their prized trout streams will dry up, and reservoirs will flood their favorite valleys and canyons. They point out, with some justification, that the west-slope drainage is already spoken for several times over on that side of the divide. All the rivers on the west side of the southern Rockies flow into the Colorado, which must provide water for the entire Southwest and part of Mexico as well. The problem is that if Tucson, Phoenix, Las Vegas and the other cities in the area were to use all the Colorado River water allotted to them by law, not a drop of moisture would reach the Mexicans. As it is, so much water is now removed from the Colorado that the river's salt content has risen drastically. Its water may soon become unusable, for with every aqueduct driven under the divide the salinity increases.

East-slope inhabitants counter with compelling arguments, most of them economic. Bounteous agricultural lands lie east of the divide in the semiarid Great Plains; but the plains need irrigation to bring forth crops. In addition, east-slope cities are growing at an astronomical rate. Each day more people need more water, both for practical day-to-day reasons and even in the pursuit of urban beauty. A Denver household-

The Mount of the Holy Cross, its distinctive emblem traced by snow, dominates a spur of the Sawatch Range in Colorado. The photograph was made in 1873 by William Henry Jackson, noted photographer of the Old West.

er who forgets to sprinkle his front lawn and lets it go brown is about as popular among his neighbors as a full-time bigamist. "Denver will not be denied water," declares one of the region's top-ranking national foresters, a thoroughly pragmatic Denverite who keeps a marvelously verdant front lawn.

And so John Carver was right. The glittery substance that covered his Shining Mountains is indeed as valuable as diamonds. And he was right in still another sense. Though no one has yet found a diamond mine in the Rockies, the quantities of other precious rocks and minerals that have been discovered here are downright staggering.

If you travel southeast from Homestake Peak to the Arkansas River, about 10 miles below the divide, you come to a town of weather-beaten Victorian quaintness called Leadville. In various gulches near town are deposits of a heavy black substance called lead carbonate, which harbors large quantities of gold and silver ore. Today the ores are mostly played out, and the town's faded brick buildings and near-empty streets seem almost bucolic. But a century ago Leadville was the noisiest, roughest, sweatiest, drunkenest—and richest—city in the Rockies. Saloons vied with brothels as the main retail enterprise. You could not walk about unarmed for fear of being murdered for your pocket money, and you ran a fair risk of catching typhoid from the open sewers, the garbage and the dead pack mules strewn about the mud streets. The danger of disease abated somewhat when the town, in a rare flourish of civic concern, built an enclosed sewer system, at absolutely no cost to taxpayers. It was financed by a $25 license fee from prostitutes. Apparently the prostitutes had no trouble paying. One of them, who went by the name of Red Stocking, supposedly retired after a single summer of hard work with a bank account of $100,000. But the real money in Leadville came from the mines. Half a billion dollars' worth of gold, silver, lead, copper and zinc has been shoveled from the surrounding gulches over the years.

Most of the ores in Leadville were formed about 35 million years ago when jets of molten rock from deep within the earth spurted up into massive deposits of limestone. Molten rock—geologists call it magma—is a wondrous substance that, as it cools, can produce all kinds of rare and precious elements. Here in Leadville some of these materials were later modified by chemical changes to produce the lead carbonate that gave the town its name, and the gold and silver ores that provided its extraordinary riches.

But there is something even more remarkable about the rocks at Leadville, and it has to do with the limestone deposits that contain the ore veins. For limestone is a sedimentary rock; and it is formed when the shells of tiny marine animals drop to the bottom of inland seas. Leadville is 10,152 feet above sea level, in the shadow of the Continental Divide. At one time it must have lain below water. What happened? Why are there massive concentrations of sedimentary rock at the very top of the mountains, and why do they snuggle, cheek by jowl, with other kinds of rock, some of them more than a billion years older? How indeed did this astonishing chaos of crests and troughs come into being, and with it the Great Divide?

The best way to answer these questions is the way the geologists do it, by looking at the rocks themselves. For the rocks contain the story of the mountains' genesis, and a geologist can read them as easily as pages in a book. His task is somewhat complicated by the fact that frequently the pages are jumbled. The mountains have risen not once but several times, and periodically have eroded away; several times over, the seas have moved in and laid down sediments, until once again the land has pushed upward to displace the sea. But a geologist, like a scholar of ancient manuscripts who pores over the scribblings and erasures of a medieval palimpsest, can piece together a narrative.

One place you can start the story is from the top, at the crest of the divide, and travel downhill into the valleys, studying the rocks along the way. Start, for instance, at Tennessee Pass on the divide near Homestake Peak, and follow the highway south. It will take you into the small, high valley of Tennessee Park, below which you will meet the Arkansas River. From there the river flows south between two high mountain ranges, and eventually cuts east into the plains of central Colorado. As you follow it downstream, you will move back in geologic time, through the history of the Rocky Mountains.

As you drive down through Tennessee Park, toward the Arkansas River, you travel from winter into spring. Snow still covers the slopes at the park's upper end, but as you lose altitude the snow dwindles and disappears. The valley itself is gently rounded in a way that indicates it has been smoothed by glaciers. And the most recent episode in the Rockies' history was indeed a shaping of the landscape by glacial ice that spread through the mountains during a succession of ice ages that began about a million years ago.

The glaciers moved down the mountains through valleys and troughs that had already been formed when—a comparatively short time ago—the region took a brief lurch skyward. The basic architecture of the Rockies was already well established; the uplift simply carried everything a bit higher. Indeed, the movement may still be going on, so slowly that its progress is scarcely measurable. But like the glaciers it also is extremely recent, having begun only about five million years ago. Five million years may seem like a long time, particularly when you consider that man himself has been around only a couple of million years. But it is a mere flicker of an eyelash compared to the august lineage of the Rockies as a whole. They started growing 70 million years ago, in a period of worldwide mountain building called the Laramide Revolution.

To gain some perspective on the Laramide Revolution, you must continue south past Tennessee Park to the Arkansas River Valley, where the highway meets the infant trickle of the Arkansas River. To your right hover the jagged pinnacles of the Sawatch Range. This is the same cordon of extremely high mountains you could see running southward from Homestake, with the Great Divide weaving through them. To your left you can see another range, the Mosquitoes. The snow has been swept clean from their massive shoulders by the prevailing winds, revealing the dull khaki of dormant tundra. Unlike the Sawatch, the Mosquitoes seem broad and rolling, like Scottish moors.

The contrast between these two ranges—the Sawatch and the Mosquitoes—is significant, for it tells you something about the character of the Laramide Revolution. When the revolution started, 70 million years ago, the land you are now looking at was awash with one of its periodic seas. Sediments were dropping to the sea bed and building up layer after layer, like thick coats of lacquer on a table top. As the sediments grew thicker the weight of the upper layers and the water itself compressed the lower layers into sedimentary rock, such as sandstone and shale. Then the earth underneath shuddered and stirred, and the mountains heaved up, carrying the sediments with them. The land continued rising, off and on, for 30 million years. This seems like a rather leisurely pace for a revolution, but the movement of mountains can be the slowest thing on earth. The forces of uplift did not work equally; there were wrenchings and shovings that contorted the landscape, lifting some areas higher than others. Sometimes breaks occurred, called faults. One such fault created the Arkansas River Valley and two mountain ranges on either side.

Two different forces worked along the fault lines in the Arkansas Valley. One was erosion. Because faulting weakens rock and breaks it apart, the land here wore away more rapidly and a depression formed. In addition, the land slipped. That is, the mountains on one side, the Sawatch to your right, started moving upward in relation to the Mosquitoes on your left. They rose so far that their very roots became exposed. These roots consist of extremely tough and ancient rocks —mostly granite, gneiss and schist—that lie below the sedimentary rocks that were formed in the inland sea that preceded the Laramide Revolution. The tops of some of the Mosquitoes are still covered with layers of sedimentary rock—thus the rolling aspect of the uplands to your left. And the craggy faces of the Sawatch consist of the more ancient rocks. They are called basement rocks, for they constituted the foundation of the entire Rocky Mountain system.

There is one area where the contrast between the basement rocks and the latter-day sedimentaries is preserved with a particular degree of clarity. Along the east slope of Colorado's Front Range, where the Rockies' easternmost slopes rise above the Great Plains, the break is exceptionally clean and dramatic. For the plains consist of vast horizontal sheets of sedimentary rock that have been building on top of one another, layer above layer, millennium upon millennium, to a depth of two and a half miles. When the Front Range pushed through them during the Laramide Revolution, it pried the sediments upward into ridges, the way a bullet, when fired through a tin can, leaves a fringe of torn metal at the point where it emerges. This region is still a two- or three-hour drive down the road and is, in fact, the destination of your journey. To reach it, you need only continue down the Arkansas River as it flows toward the Great Plains.

The air warms steadily as the Arkansas Valley drops in elevation. The scattered pine groves on either side of the road have changed character. The lofty stands of lodgepoles and ponderosas higher up have given way to a scrubbier tree with gnarled branches and a squat, bushy profile. These are piñon pines, and their presence signals a transition between the cool, moist climate of the mountains and the dryness of the plains, where you are now headed. Sagebrush fills the air with a fine, spice-shelf fragrance.

The river bottom widens into tableland, the last pine groves disappear and the Mosquitoes to the left dwindle into foothills. But the Sawatch towers strong as ever to the right, the divide with it. The river passes a coterie of peaks named by chauvinistic climbers from various Eastern universities: Mt. Yale and Mt. Columbia, broad and sturdy; Mt.

The Arkansas River churns at the bottom of the Royal Gorge, the 1,000-foot-deep cut in the pink granite of Fremont Peak in Colorado.

Harvard, the Rockies' third highest, which a climbing party from that institution once tried to convert into number one by adding a 20-foot aluminum pole to its summit; and Mt. Princeton, a neatly symmetrical triangle rising prim and independent from the valley floor.

Here you should allow yourself a detour. For among the peaks behind Mt. Princeton, buried in veins of granite and smoky quartz, is an exceptionally beautiful and interesting mineral. It is aquamarine, which comes in hexagonal prisms sometimes measuring as much as seven inches long. Clear as glass, ranging in color from palest green to icy blue, these prisms might as well be chunks of the Pacific, silent and frozen, transplanted to the mountaintops.

All through the Rocky Mountains you can find stones of similar brilliance and fascination. You cannot walk along some stretches of the divide or down the rivers and through the canyons, without filling your pockets with them: mica, quartz, garnet, agate, amethyst, turquoise, a rock-hunter's treasure chest open there under your feet. The search for beautiful stones is called rock hounding, and you can pursue it as you continue down the Arkansas River to the Great Plains, sifting through the history of the Rocky Mountains.

The south end of the Arkansas Valley is blocked by the snow-bound peaks of a range called the Sangre de Cristo. The river turns east to avoid them and begins its cut through the foothills of the Front Range, the last mountain barrier to the plains.

The Grand Canyon of the Arkansas River begins quietly, with little fanfare and no hint of the splendor to come. The river dips between some low hills, you hear rapids, the air grows suddenly damp. Then the hills fall away and you drive through fields again. The sequence repeats itself: a drop between hills, the noise of rapids, and the land flattening out. But with each cycle the hills close in steeper, the waters roar louder and soon the fields disappear entirely. You find yourself driving down a corridor between two rock walls, their tops bright with sun, their bases awash in shadow. At one point a wall of limestone tilts above the road, white as a snowbank: a former sea bed. Farther on, a seam of coal where once there had been a swamp. Then the river plunges steeply, and the road leaves it to climb through the bluffs to the rolling uplands above. You approach the mightiest cut of all, the Royal Gorge of the Arkansas, from above.

A confirmed rock hound will stop before he reaches the gorge and sift the ground for gems. For there is beryl to be found at a dozen or so mineral sites in the area, in shades of yellow, white and green, with the texture of porcelain. You may hunt for tourmaline, with its lozenge-shaped crystals, black or crimson, arranged in sunbursts that have been known to reach four feet in diameter. You may discover agates or blood-red garnets of sufficient clarity and brilliance to be set into necklaces and rings. Or you may find only rose quartz, as common here as sea shells on a beach. Yet even this ordinary stone, a luscious strawberry-ice-cream pink, takes on a special brightness when you manage to discover it yourself.

Only then, your knapsack heavy with stones, will you approach the lip of the gorge. You would be advised to leave your car by the enclave of curio shops at the entrance to a small commercially run park, where a suspension bridge has been strung across the gap, and walk off by yourself through the yucca and cactus toward the rim. Suddenly the land will stop and your boot will hover, midstep, over—nothing. A vertical slice of the earth's crust, several hundred feet wide and more than a thousand feet deep, has simply been annihilated. The cliffs plunge clean and sheer to the river bed. Far down, scarcely noticeable at the bottom of the pink granite cliffs, the river has been diminished to a tiny etched line of silver.

An illusion of course, for by this point the Arkansas is not tiny at all. On its way down from the divide near Tennessee Pass it has gathered the meltwater from the snow-covered mountain ranges upstream and grown into a rampaging torrent. You can descend to it by means of a cable car. The sensation as you ride into the abyss works an odd transformation on your own sense of importance. Imagine yourself reduced to microbe size, lowered into a hairline crack in one of the rose-quartz crystals in your knapsack. You seem to penetrate into the heart of a single immense pink stone. And in a sense you are doing just that. The Royal Gorge slices into a unified formation of pink granite that reaches all the way to the summit of Pikes Peak, 20 miles to the north. Geologists call this structure the Pikes Peak batholith, a domelike swelling of basement rock that bowed upward during the Laramide Revolution, was eroded down, and about five million years ago started rising once again. The river cut down through the batholith at the same leisurely rate that the land was moving upward—at times no more than a foot every 2,500 years—thus creating the Royal Gorge.

You must resist the temptation to linger on the thin ledge of riverbank at the bottom of the gorge, with the grind and tumble of the rapids echoing between the pink crystalline walls, the light shafting down from the sliver of blue sky a thousand feet overhead. For the purpose of this trip is to reach the plains, where the river cuts through other rock formations, less spectacular in a way, but equally fascinating. And the plains, from this point, are only five miles away.

From the top of the Royal Gorge, the highway glides eastward across the uplands, then drops about 1,500 feet through timbered foothills and emerges onto the edge of the plains. Behind you towers the Pikes Peak batholith—basement rock of the Front Range mountains. In front of you is a series of parallel ridges consisting of sedimentary rock. Each ridge is about 250 feet high, running north and south along the edge of the mountains as far as the eye can see. Not a tree grows anywhere, and only the scantiest of grass covers. The colors of the ridges are muted but quite distinct: red in the ridge closest to you, then tans, browns and grays in the ridge just beyond. A side road leads to the crest of this second ridge, and when you reach the top of it you see still other ridges to the east, all lined up parallel, rolling in across the plains. These are the exposed edges of the Great Plains sediments, torn upward by the rock mass containing the Pikes Peak batholith.

To a geologist each sediment tells a story. The first red-colored ridge is part of something called the Fountain formation, and it testifies to the rise and subsequent decay of an ancient mountain range that preceded today's Rockies. These Ancestral Rockies, as they are called, elbowed their way upward about 325 million years ago, and 175 million years later they were gone, eroded flat. The Fountain formation is one of the products of that long period of erosion. As the mountains disintegrated, sand and gravel washed out onto the plains at their feet, and under the weight of later sediments were compacted into a tough sandstone; the red color comes from iron oxide, one of the ingredients in the chemical cement that holds the sandstone together. All along the edge of the plains, at the Garden of the Gods in Colorado Springs, at the Park of the Red Rocks in Denver, in vertical sandstone slabs behind Boulder, Colorado, the Fountain formation was tilted on end when the Front Range pushed through it 70 million years ago.

The mountains' reach for eminence also pried up the next ridge to the east, where you now stand. Its bulk is a series of deposits called the Morrison formation, laid down about 140 million years ago. The area was then in a swampy transitional period between sea and land. Giant dinosaurs lumbered through these wet lowlands. In 1877 a geologist named Arthur Lakes, digging in the Morrison formation near Denver, unearthed the fossilized bones of a 70-foot brontosaurus. Since then no deposit on earth has yielded such a trove of dinosaur skeletons as the Morrison formation.

The Morrison formation is mostly shale, a soft and crumbly rock, and under most circumstances it would have worn away without leaving a ridge. But on top of it rests a layer of extremely tough rock called Dakota sandstone, which has protected it. Dakota sandstone is the remnant of an ancient beach that edged inland across the Morrison sediments about 135 million years ago as the sea crept over them from the east. If you pry a chunk loose with your hands—it comes apart in flat layers—you will see ripple marks left on its surface by the tides. If you are lucky you may happen upon the fossilized shell of a 130-million-year-old clam. Dakota sandstone is a porous material, and it sops up water and carries it miles away under the plains. Farmers in Nebraska drill down to it when they need well water. For the Dakota Hogback, as this ridge is called, is the merest frayed edge of a vast sheet of sedimentary rock that spreads out for millions of square miles under the surface of the plains—the legacy of the vast inland sea that covered the area of today's Great Divide.

And so it is with all the ridges. In your mind's eye, as you stand on the crest of the Dakota Hogback, they seem to roll in across half a continent to break like surf against the roots of the mountains. This was the landscape, built up through eons of flooding by the sea, before the Rockies broke through to create the brief episode of the Great Divide. Far to the east, in a thin brown haze at the edge of sight, the land merges imperceptibly with the sky.

NATURE WALK / # Along a Glacial Stream

TEXT AND PHOTOGRAPHS BY DAVID CAVAGNARO

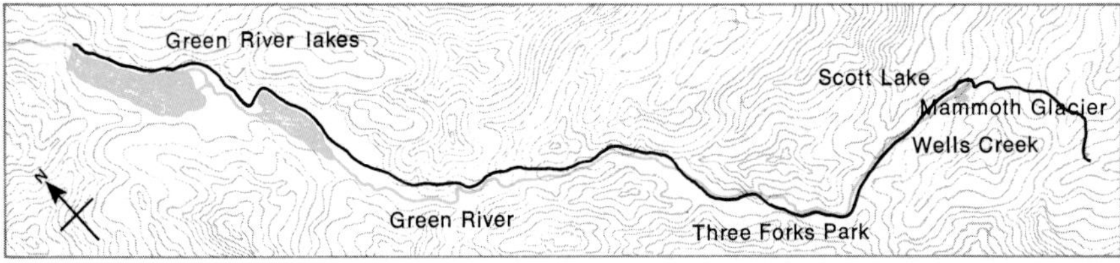

ROUTE OF THE HIKE FROM THE GLACIER

On a September day photographer David Cavagnaro and two companions reached the top of 12,000-foot-high Mammoth Glacier, near the Great Divide in the isolated Wind River Range in Wyoming. Mammoth is in the heart of the seldom-visited Bridger Wilderness Area, and its melting ice is a source of the Green River, which flows down the continental watershed's western slope and runs eventually into the great Colorado and thence to the Gulf of California. The hikers proposed to follow the river from its headwaters at Mammoth down through its upper valley to the two Green River lakes, where the river flows out of the mountains. The trip took four days and Cavagnaro kept a journal, excerpts of which follow.

Saturday: Mammoth Glacier fills an immense cirque, or basin, about two square miles in area. It is surrounded by saw-toothed peaks more than 13,000 feet high, including Wyoming's highest, Gannett Peak; all of them had fresh snow from recent storms. The view in every direction was fantastic. Far off toward the southwest we could see the long ranges that mark the borderland between Wyoming, Utah and Idaho. Far to the northwest the entire skyline of the Teton Range was visible. From a nearby col we looked down into the deep canyon of the upper Green River valley that we would be passing through during the next few days of hiking; Squaretop Mountain, a landmark that is toweringly visible from one end of the canyon to the other, lay far below us.

Jon Braun and Vince Lee went off to hike around nearby Minor Glacier, and I was left to myself on an icy wind-swept pass between two towering peaks, about 450 feet above the slope of Mammoth. Though the day was clear, the west wind that blew through the pass was so frigid that I

SNOW FIELD ABOVE MAMMOTH GLACIER

had to keep moving to keep warm. Exploring the snow field above the glacier, I found the tracks of a coyote and of two mountain sheep bound through the pass on some errands of their own. I followed the tracks for a while until the blowing snow obliterated them.

By then it was getting well on into

THE GLACIER'S ICY EDGE

the afternoon, and I began descending the snow field, half walking, half sliding on its frozen surface. Blasts of wind scoured the glacier and raised great swirls of new, powdery snow all around me, veiling the low sun and cloaking the scene in magic.

The bottom, or snout, of the glacier is only 100 yards or so across and drops steeply into a shallow gravelly basin just below the rim of the cirque, as though the vast bowl of the cirque had been tipped slightly and the thick, plastic contents had spilled over the downward edge. I walked to the very tip of the snout. From under its thin icy edge flowed a burbling stream of water about three feet wide—the beginning of

Wells Creek, the principal headwater of the Green River.

My companions and I had agreed to rejoin forces and spend the night at the east end of Scott Lake, a tarn fed by the runoff waters of Mammoth and Minor glaciers. It lay about two miles below me and was already in shadow as I left Mammoth's snout. I could just make out Jon and Vince setting up camp there. Above me the late-afternoon sun was turning the peaks a rich scarlet, and at my feet a bright golden glow was reflected in the first meanders of Wells Creek just 100 feet down the gravel bed from its glacial source.

WELLS CREEK AT SUNSET

I hastened to join the others and got there just as darkness fell. A welcome fire was burning and dinner was in the making. Not much later we turned in, exhausted.

Sunday: At sunrise I set out in the growing light for a walk around Scott Lake. It is a remarkable body of water. It fills a steep-walled basin into which the melting snow and ice of both Minor and Mammoth glaciers flow. Because of the very active nature of these glaciers, their runoff water is heavily laden with rock flour—powder-fine particles of rocks that have been ground up by the glaciers' ponderous advance. Carried in suspension, the powder makes the water an arresting milky blue.

The lake is rimmed by a sparse selection of alpine vegetation, most of it low growing in deference to the inhospitable weather at this altitude. Two species of huckleberry covered the ground here and there; poking up in one stony place were several blooms of rosecrown, which is also called stonecrop because it can grow in very thin, rocky soil. In fact rosecrown, which looks deceptively like a red clover blossom, is of a family of succulent plants that are often cultivated in rock gardens.

At another spot, where the ground was damp, I found the blossoms of two other flowers nodding at each other in the morning breeze—two plants of delightful aspect but uncharming names: lungwort and fleabane. Lungwort, a kind of bluebell, got its name because it was once supposed to have a healing effect on certain respiratory ailments. It is a treat for grazing deer and elk, but whether it does their lungs any good I have no idea.

Fleabane, which looks like a lavender daisy or aster, also has a place in folk medicine: it was thought to drive away fleas, and to that end the dried, powdered flowers were once sprinkled in dog kennels. The idea seemed farfetched up here on the shores of Scott Lake at 10,480 feet, but the delicate flowers were nonetheless decorative.

As I finished my exploration of the lake a stiff, cold wind was beginning to blow and white puffy clouds began racing across the sky from the west. Change of weather for sure. We had to make a decision: stay another night at the shore of the lake and chance bad weather tomorrow, or leave that afternoon.

It was a serious decision because the only way out is a hazard even in good times. The lake's outlet is a narrow crack in the bank at the western end through which must pass all the water that has collected from a glacial watershed of several square

GLACIER-FED SCOTT LAKE

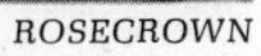

ROSECROWN

LUNGWORT

FLEABANE

RUNOFF FROM SCOTT LAKE

miles. The crack leads into a rock-strewn chute that drops 680 feet in one quarter of a mile and is never more than 50 feet wide at its base, sometimes as narrow as 20 feet.

Wells Creek goes down this chute on its way to join the Green River, and we planned to go that way too; but with its bouldered stream bed and steep pitch, the chute is a very dangerous route, and should the rocks be slick with rain, ice or snow it would be virtually impassable. We took another look at the streaming clouds and decided to tackle the chute right away.

We had waited almost too long. When we reached the top of the chute, we found the wind raging, funneled through this narrow bottleneck with the strength of a full gale. Negotiating the chute was like de-

FROM THE TOP OF THE CHUTE

scending a huge shaky ladder whose rungs cannot be trusted. Much loose material and slippery algae from the high water of Scott Lake's springtime overflow made for a precarious descent at best, and with our packs on we were constantly thrown off balance by blasts of wind.

From the bottom of the chute the terrain fans out into a wide, steep rocky slope leading down to a place called Three Forks Park, where Wells Creek joins two other streams to form the upper Green River. The slope owes its desolate aspect to a cataclysmic washout of Scott Lake that occurred sometime in the last half century, though no one knows quite when. The top of the chute had evidently been dammed at one time by a rockslide that backed Scott Lake up so that its surface was perhaps 20 feet higher than it now is. Some of the sheer granite walls around the lake still show the marks of this high water level.

Then apparently this natural dam burst, more or less all at once, sending a great percentage of the former Scott Lake down the chute in one huge washout. That must have been a spectacular show—one to have seen from a considerable distance.

The force of the washout had carried the boulders from the dam down the chute and deposited them randomly on the slope below it. Now a few trees, mostly neat symmetrical firs up to 20 feet tall, have grown to relieve the barren landscape. Small gardens of fruit and flowers have formed in stony recesses and we feasted on the last of the season's red mountain raspberries.

We decided to camp in a place about halfway down this boulder field, near the creek and overlooking a clump of quaking aspens blazing yellow in the late-afternoon light. It was a place meant more for elves than for people; someone, man or elf, had deposited there a great flat stone that we used as a table. It made a perfect place to camp.

Monday: We planned today to hike the 12-mile length of the upper Green River valley, so we got a dawn start despite an intermittent rain that plagued us the first hour or so. From Three Forks Park just below our campsite, the river commences an indolent meander down the valley. Though cradled by steep-sloped mountains, the valley floor is almost level here, and the slow-paced river has deposited rich sediments along its course. In these wet, marshy places a lush growth of sedge and grass carpets the floor from wall to wall. A month earlier the carpet would have been a rich green. Now, after several nights of

SUBALPINE FIR IN A BOULDER FIELD

A CLUMP OF GOLDEN ASPEN

MARSH GRASS ON THE GREEN RIVER

heavy frost, some of the grass had turned to a yellow-brown fall color that contrasted starkly with the somber rocks on the mountains beyond.

To avoid having to slog through marsh we walked through the lodgepole pines and Douglas firs that maintain a sturdy foothold on the stony hillsides. The pines harbored in their undergrowth a variety of small plants decked in autumnal hues. My eye was caught by a plant none of us could identify. It had small oval leaves, bright orange, with curious round holes in them. Jon explained the holes as the work of leaf-cutting bees, which chew out circles of leaf almost as perfect as those a store-bought perforating gadget would make. The bees fly the round cutouts back to their nests and use them to close off the cells in which they have laid their eggs; conveniently, they always chew out circles of leaf one size bigger than the cell openings.

A bit farther on I noticed the red leaf of a wild geranium, cousin to those in window boxes everywhere, and then two more plants, growing side by side, whose leaves were much the same orange-red but of very different shape. The dandelion leaf was long, narrow and deeply notched, the strawberry's more like an outsized clover leaf. These two were lucky survivors, for both plants are favorite food for all manner of creatures that range the forests in summer, from grouse to black bears.

After several hours hiking through the forest we decided to get another perspective on the river and its val-

LEAF HOLES CHEWED BY BEES

WILD GERANIUM LEAVES

DANDELION AND STRAWBERRY LEAVES

GREEN RIVER VALLEY AND LAKES

ley. So we struck off for a high ridge at the top of a rocky slope. On the way up we came suddenly upon an elk boneyard, the remains of about eight carcasses, all apparently of cows and calves, much chewed by coyotes and scattered about. The previous winter had been a hard one in northwestern Wyoming and the elk had suffered; evidently these animals had foundered in a very deep snowdrift and, unable to move or find food, had perished together.

As we trudged on, a chill wind reminded us that the countryside and its animals would soon enough have to cope with winter again; but the scene that greeted us as we got to the top of the ridge was warm with autumn gold. Below us glowed the amber marshes where the river wandered; off to the west we could see both the upper and lower Green River lakes and faintly in the distance the broad lower valley through which the Green flows after it leaves the mountains. Over all, scattered dark clouds hurried raggedly along, casting restless patterns of light and shadow on the land.

Among the dodging shadows on the surface of the upper lake were two black specks that we realized were animals swimming across. We guessed they were moose, though at that distance it was hard to tell. We found out a bit later, though, when, coming down the hill, we saw two creatures grazing in the marsh, still shaking the water from their coats. They were out of camera range but quite identifiable: a cow moose and her full-grown calf.

LOWER LAKE AND SQUARETOP MOUNTAIN IN THE MIST

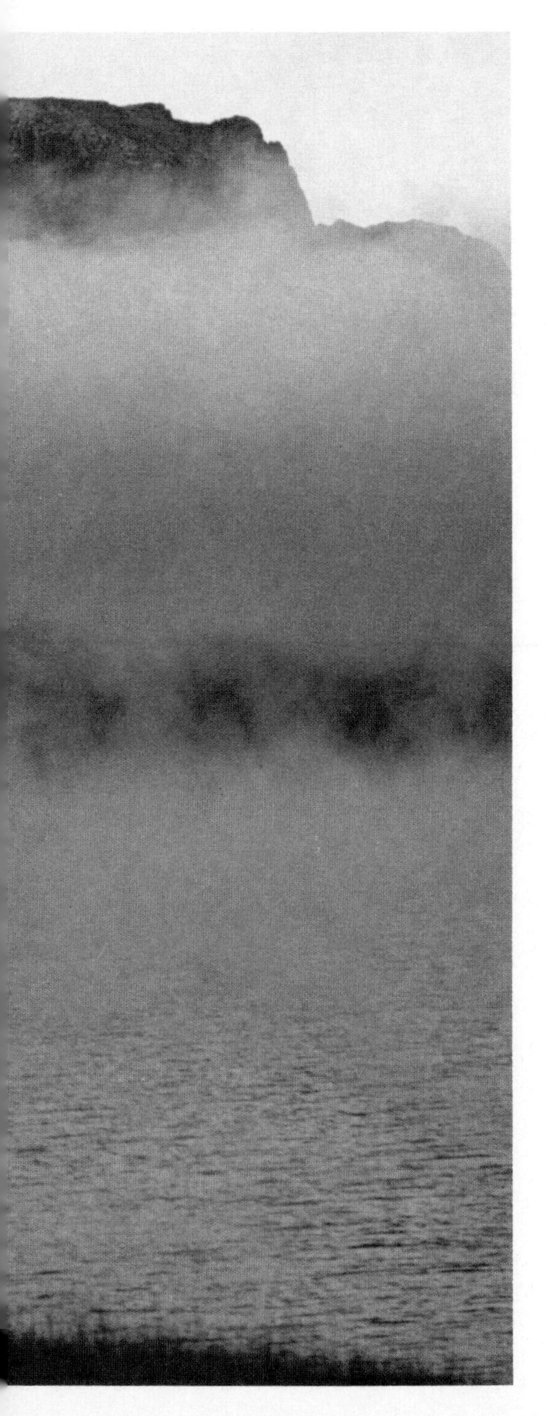

By now the chill in the wind really meant business, and we hastened to the lake where we would camp for our last night in the wild.

Tuesday: Morning arrived clear and very cold. Frost fringed every leaf, glistened on my camera cases, crackled on the covers of our sleeping bags. Here and there I found a favorite sight of mine: "frozen mouse breath," formed during the night when the warm breath of rodents in their burrows rose and condensed as frost on the grass blades clustered around the nest entrances.

The fire we got going immediately was much needed, but while breakfast was cooking I deserted its warmth to take a quick walk part way around the lower lake. The relatively warm water was sending up clouds of mist in the frigid morning air, establishing an eerie setting for the view across the lake toward the way we had come. Hulking darkly above the mist and the steely water was old Squaretop, which I had looked down on three days ago from Mammoth Glacier. At my feet were more frostbitten plants: a few pale blue gentians haloed with rime, and some green and red cinquefoil plants, ice-rimmed and posed elegantly against a lichened rock.

Over breakfast we talked wonderingly of the astonishing concert we had heard in the small hours of the night before. Sleeping lightly because of the cold and the light of a brilliant full moon, we had been awakened sometime after midnight by an amazing sound floating out of the silence, a sound one dreams

FROZEN MOUSE BREATH

CINQUEFOIL AND LICHEN

RIMED BLUE GENTIAN

about but never really expects to hear. It was the bugling of a bull elk, coming from the slope above us. We listened in awe as the bull slowly drew closer to camp, sounding his call every few minutes. Soon we heard another bull return the call from far up the canyon, and the two began to exchange announcements. Then, as though this were not enough for one night's symphony, coyotes began barking, alternating with the bulls in a spooky chorus. At dawn the calls of coyotes and elk were still emanating faintly from the far reaches of the canyon, and to these distant sounds a new call was suddenly added—the whistle screech of a great horned owl perched in the forest across the lake. It was the greatest natural chorus I have ever heard.

As we finished breakfast the sun burst over the ridge, the mist gradually dissipated, the frost slowly melted. We packed our gear and hiked to the end of the lower lake. We took a last look at its pristine expanse, now blue and sparkling in contrast to its gray early-morning aspect, and then regretfully turned and headed for the road where we were to be met by friends in a car.

The land here was much altered. This was open, gentle country, and the river, wider now, swept in a big double bend through brilliant stands of golden aspen and clumps of pale green, fragrant sagebrush. The sage does not turn in the fall, and its low pastel bushes were set off by the yellowing leaves of a variety of other seasonal plants.

We walked lightly, exhilarated by our experience in the mountain country and grateful for having been treated to so many of its moods in only four days.

GREEN SAGEBRUSH AND AUTUMN LEAVES

SAGEBRUSH AND ASPEN ALONG THE LOWER GREEN RIVER

3/ Adventure in the San Juans

Mountains are to the rest of the body of the earth what violent muscular action is to the body of man. JOHN RUSKIN/ *MODERN PAINTERS*

Mountains, for some people, are magnetic. Certain individuals, when faced with a rise in the landscape, feel an odd compulsion to start climbing. And so they strike out across hogbacks and foothills, through canyons and spruce forests, up past timberline and out across the alpine tundra, until they meet the bare stone faces of the peaks themselves. Here most of us, appropriately humbled, would tip our hats and turn downslope. But for the ultimate mountain man this is not enough. He must climb farther, up those grim rock faces, battling his way across the wrinkles and creases until he stands at the very top, higher than anything—or anyone—else in sight. The tougher the ascent, the better he likes it.

You may quickly cross my name from your list of ultimate mountain men. High-altitude gymnastics cause butterflies to flock in the pit of my stomach. It was therefore something of a paradox that I found myself, on a summer weekend not long ago, attached to a party of climbers from the Colorado Mountain Club as they headed into the San Juan Mountains in southwestern Colorado. Their objective was a group of peaks called the Grenadiers, well over 13,000 feet high and dear to the hearts of mountaineers, who consider them particularly challenging.

And indeed few mountains in all the United States offer more climbing excitement than the San Juans. They heave and toss with every conceivable type of land formation—uptilted blocks of sedimentary rock, volcanic crests, granite steeples, breakneck gorges and slender matterhorns sharpened by glaciers. They are high mountains, with almost a dozen peaks breaking 14,000 feet. They can be dangerous mountains, for certain pinnacles are so severely weathered that the rock peels away when touched. The temptation to tie into a climbing rope and head for their summits is—for some people—irresistible.

My own purpose in coming to the San Juans was slightly different. I had learned by studying a topographical map that from one of those summits I might look out over an exceptionally beautiful section of the Great Divide. I was drawn, too, by the promise of solitude. These mountains comprise the largest single range in the United States Rockies. Counting all its offshoots and subsidiaries, the San Juan massif encompasses more than 10,000 square miles, space enough to enclose the state of Vermont, with a sliver of New Hampshire besides. Yet almost no one seems to live there; the map shows fewer than a score of towns, plus a light salting of mining sites, most now abandoned.

The divide sweeps through this unmolested countryside in a prodigious hairpin curve. Coming off the southern foothills of the Sawatch Range, it veers southwest for a distance of about 80 miles to the apex of the curve. Here it doubles back again, running southeast for another 80 miles toward the New Mexico border, where it leaves the Rocky Mountains for good to lose itself in the sandy scrub-covered hills of southwestern New Mexico.

Cupped inside the bend of the hairpin is the source of a great American river—the Rio Grande, which rises on the curve's apex and sweeps 1,885 miles to the Gulf of Mexico. The Grenadiers thrust upward just west of the hairpin, about three miles from the divide, directly over a ridge from the headwaters of the Rio Grande. This is the tumultuous heart of the San Juan range, a land so topsy-turvy that not a single acre is suitable for farming.

Rugged mountains are usually beautiful mountains. The San Juans provide more beauty than a person reasonably needs—all those convoluted shapes, those chromium colors, those humbling distances, those obliterating tonnages of earth and rock. Amid such grandiose perspectives a person's sense of scale breaks down. Like Alice journeying down the rabbit hole on her way to Wonderland, the body forgets what size it is. Each time you round a rock buttress, or step from a forest into the dazzle of an open meadow, dimensions spread out that just should not be there. The scenes shift too quickly. Beyond a stand of spruces, suddenly and for no immediate reason, the earth pitches down for a third of a mile before it stops. A river boils along the bottom, with dwarf cedars and yuccas—desert plants—on its bank. Across the canyon and up through the forest, the land heaves above timberline, where fields of glacier lilies bloom in the wakes of retreating snowbanks.

The foundation of the San Juans' beauty is their geology. Much of the rock in the western part of the range is sedimentary and fairly young as geologic time goes—perhaps no more than 150 million years. Its colors exhaust the spectrum: sandstone and shale in vivid laminations of red, ivory, buff, saffron, charcoal, lavender, mauve. One summit, called Red Mountain, looks as though it were composed entirely of red and white pigment suitable for turning into house paint.

The Grenadiers themselves stand in the center of the range, where the strata are nearly a billion years old, and long since reformed by the pressures and heat of earth building into metamorphic rock. These ancient formations have been deeply eroded and battered by the elements until they have taken on the pink-gray color of weathered brick. "Steep faces and ridges of a hard quartzite which makes for sporty climbing," is the way they are defined—with alarming relish—in the *Guide to the Colorado Mountains,* the CMC's family bible.

In the eastern half of the range the mountains occasionally have a charred look, as though they had been licked by flame. In a sense they have been. During the Laramide revolution, 70 million years ago, fountains of lava bubbled up through the fissures in the overlying sediments and washed out across the land. Remnants of these lava flows crown many of the eastern mountaintops.

Landscape is geology with plants on it, an observation first made by a geologist, and one that applies with particular force to the San Juans. For in these mountains, plants provide the most beguiling spectacle of all. During the brief alpine summer, when the snow leaves the upland meadows, a wild flourish of alpine flowers bursts through the soil: crimson of king's-crown and paintbrush, gold of buttercup and avens, pinks and lavenders of penstemon and fleabane, blue of harebell and columbine. Lower down the forest takes over, spruces, firs and aspens closing out the sky with a mottled green umbrella. Daisy, raspberry, grape, twinberry, larkspur spring up where sun breaks through the holes.

Perhaps because the plains to the south and east are particularly arid, these forests seem unusually verdant. The San Juans capture most of the rainfall in southwestern Colorado, leaving the outlying countryside semiarid. Parts of the San Luis Valley, which lies in their rain shadow to the east, are hummocked with gigantic sand dunes, some of them 700 feet high. These are the tallest dunes in the country, and Congress has set them aside as a national monument.

Part of the charm of the San Juan landscape is historical. The hillsides are crisscrossed by old mining roads and pitted with tumble-down mine shafts. The San Juan county seat, about 10 miles north of the Grenadiers, began as a mining town. It is Silverton, founded in 1874, elevation 9,302 feet, winter population 780. The first settlers moved there in 1871, after someone discovered gold in a nearby gulch, which implies that Silverton was misnamed. But the gold played out, someone else found silver, and the settlers began hacking it out of the mountainside at the rate of about $100,000 a year.

A modest return, considering the price. The San Juans in 1871 were so remote that the first maps had not yet been completed. The only trail into Silverton, in summer, was a 50-mile haul over the Continental Divide to the east. In winter there was no way in. For six months at a stretch, snow blocked the passes and Silverton was as isolated as an arctic weather station. No outside newspaper, no fresh vegetables, no maggot-free flour, no new faces from November through April. The year's wildest excitement—Christmas, Easter and the Fourth of July rolled into one—was the arrival of the first pack train in spring.

In 1876 the pack mules appeared on May 2, and the Silverton newspaper just about lost its head with jubilation: "Somebody gave a shout, 'turn out, the jacks are coming,' and sure enough there were the patient homely little fellows filing down the trail. Cheer after cheer was given, gladness prevailed all around, and the national flag was run up at the post office." Civilization reached closer in 1882, with the arrival of the Denver and Rio Grande Railroad, a slim ribbon of track climbing up the Animas River from Durango, 46 miles to the south. Now it took only half a day to get to Silverton, or out of it.

The train is still the most interesting—and practical—route into Silverton and the interior of the San Juan Mountains. The D&RG is the kind of nostalgic relic, from an era when trains were called iron horses, that brings a warm glow when mentioned among connoisseurs of railroad history. Its narrow-gauge tracks are laid only a yard apart (as opposed to four feet eight and a half inches for standard gauge), which makes it easier to thread them through canyons and up mountainsides. Even so, the D&RG tracks have been washed out by floods, pounded by rock slides, shoved about by avalanches, and most stretches have been replaced many times over.

Hulking 12,890-foot-high Red Mountain is one of a trio of identically named peaks in the San Juan Range in southern Colorado. The

three are so called because their rocks and soil contain iron that has oxidized on exposure to the atmosphere and turned rust colored.

In a few places, where some of the original track survives, the train rides on silver. The local iron used in forging them contained a tiny percentage of silver ore. Along these tracks the traveler is drawn by an ancient puff-and-chug, coal-burning steam engine. Every 15 miles or so the engine has to stop to take on water for its boilers, for the climb is a thirsty one. It rises almost 3,000 feet from Durango at the edge of the plains into the mountains at Silverton.

No roads lead to the trailhead into the Grenadiers, and so climbers invariably ride the train. You fumble aboard at Durango with your backpack and climbing boots, and find a seat in one of the antique cars. You watch the countryside roll by: the fertile bottom land of the Animas Valley flanked by 1,000-foot cliffs with rust and ivory faces. Cattails and willows grow along the riverbank, bayonet clumps of yucca in the drier spots, barricades of thistles in purple bloom, regiments of plains sunflower, each yellow disk tipped toward the sun at exactly the same angle as the flowers that flank it. The early settlers sometimes harvested these various plants for food, grinding the starchy cattail roots into flour, eating sunflower seeds, and washing, in emergencies, with a soapy secretion given off by yucca root.

The train pulls higher, into a forest of ponderosa pine and scrubby Gambel oak. Trees shut out the valley walls and the eye turns downward. The train has climbed the pink granite rim of the Animas Canyon and now seems to hover directly above the river bed. Four hundred feet below, the water seethes and crashes with a sound that can be heard above the chugging of the engine. The river's full title is Río de las Animas Perdidas, which is Spanish for "River of Lost Souls." At least a century before the first Yankee prospector pounded his claim stakes into a San Juan mountainside, Spanish adventurers from Santa Fe had snooped about the foothills. No one knows exactly which hidalgo discovered the Animas, but one legend says that five of the king's soldiers wandered up it and never returned, thus providing the River of Lost Souls with a name. Looking down into the cauldron of white water from the train window, it seems entirely possible.

To get into the real climbing country of the Grenadiers, you leave the train about 10 miles south of Silverton. The forest opens, the land temporarily flattens and the engineer lets you out in a grassy meadow walled in by high peaks. You walk back down the tracks, along the west bank of the Animas, to the trailhead. There are wild raspberries to eat on the way, if the month is right, and wild flowers enough to keep you riffling the pages of your field guide. The goldenrod blooms thick in midsummer, as do the magenta spikes of fireweed—so named because it is one of the first plants to seed itself after a burn. A small flower with five delicate pink petals turns out to be a Frémont geranium, a wild cousin of the hydrant-red blossom that people cultivate in window boxes.

Scores of daisy-like composites bloom along the track. They have yellow centers and pink or lavender petals, and only a botanist can tell you what species of daisy or fleabane they belong to. These look-alikes can be found at all elevations, and they share in the same bit of botanical deception. What seems to be a single flower is actually hundreds, all grouped together in a tight bunch. Each lavender petal, and each bit of yellow fuzz in the central disk, is a tiny individual floret with its own pistil and stamen. Easier to identify is another composite called pearly everlasting, which looks and behaves exactly like its name. The round white flower heads do indeed resemble pearls; and in fall, long after the first frost, they hold together when other flowers have withered and dropped away. Because of this remarkable staying power, commercial florists grow them to preserve and sell as strawflowers. They are dried and sprayed with bright colors, a process that would presumably make them garnet or turquoise everlastings.

The trail begins in a forest of aspen and white fir at a point where a tributary stream flows into the Animas from the east. It follows the stream to the source, Balsam Lake, the most appropriate campsite for a party climbing the Grenadiers. The stream is called Tenmile Creek, a name that fortunately does not refer to its length; it only seems that long. The *Guide to the Colorado Mountains,* which talks of gut-wrenching precipices with a breezy equanimity, has this to say about the trek into the Grenadiers: "Tenmile Creek, 10 miles down the river from Silverton, climbs 2,800 feet in less than five miles to Balsam Lake. The route starts with the drowning of all but the fittest—a crossing of the Animas that is impossible early in the year—and continues with a line of march which has only occasional resemblance to a trail."

The river can be managed by stringing a rope across it. But the trail is another matter, particularly for someone whose muscles—and will power—have been eroded by the normal amount of sedentary living. The hike through the aspens is pleasant enough. The sun filters through the apple-green leaves like the glow through a stained-glass window. Moss turns the rocks underfoot into soft green hummocks, and tufts of what appears to be hair, a grayish lichen called old-man's-beard, hang from some of the lower branches. Both require a lot of water, thereby indicating a degree of moisture unusual in the Rockies.

The trail grows steep, then steeper, climbing down gullies and up them, across bizarrely tilted meadows, past the aspens and white firs and into the somber shadows of Engelmann spruces. Here it gives up entirely, out of sheer exhaustion. The heavy spruce boughs close out the sunlight. Windfall tree trunks, a yard in diameter, barricade the forest floor. They must be climbed over or walked around, which is a good way to encourage frustration and lose all sense of direction. These spruces are a climax stand that has grown undisturbed for half a millennium and clearly has no patience with intruders.

It has taken four hours to climb from the Animas River to the interior of the spruce forest, a distance of about four miles. Four hours straight uphill, until each step has become an agonized wrench. The straps of the backpack cut into shoulders and hipbones, the ankles wobble like loose piston rods. The thigh muscles quiver and complain, the throat grows dry and grainy, and the breath rasps. Of all the compulsions that draw people into the mountains, one is surely a predilection for self-torture.

I had lagged behind my fellow climbers and was hiking alone. Suddenly I heard a large animal flailing through the trees behind me. I stopped, but the animal did not. When it came within sighting distance it gave a shout. "Hi, there," it said, and turned out to be a youthful hiker with a heavy carton tied to his back. An advance party of the Colorado Mountain Club had set up camp a few days earlier on Balsam Lake, and this lad had hiked back down to the river to haul in additional supplies.

"If this is your first day out, you've got another two hours to the lake," he told me. "It'll be dark in an hour. You could camp here and hike in tomorrow morning."

I told him thanks, I'd think about it, and begged a packet of peanut-butter crackers for dinner. He bounded off through the timber with a degree of energy that seemed, to someone in my present condition, positively nauseating. I trudged along for another half hour, the air turning cool in the approaching dusk, until I reached a bluff overlooking Tenmile Creek. Across the creek bottom the valley wall lifted in immense crenelated dumplings of barren rock, glowing violet and rose in the declining sun. It seemed like a good spot to quit. I spread my sleeping bag under a spruce and watched the twilight flood the sky like a soft gray liquid.

A white-tailed ptarmigan hen, in her mottled summer plumage, is well camouflaged against the alpine terrain she inhabits along the divide. When winter's snow falls, her protective coloring will change to pure white.

There is a certain excitement to waking up in the mountains with only sky overhead. The air braces one like ice water. The senses sharpen, the brain starts awake, and the juices flow with rare and gratifying vigor. All my weariness of the night before had drained away. The remains of a chocolate bar and a cupful of water from Tenmile Creek set me up for the hike to Balsam Lake.

Bird cries rippled through the conifers: the musical *whee-twee* of an Audubon's warbler and the *klak-klak* of a Steller's jay. This latter bird has the bearing of an aristocrat, with marvelous sapphire plumage and ebony crest, and it glides regally from branch to branch. But the effect is ruined by its voice, which is raucous as a fishwife's. At one point, with a furious thumping of wings, a blue grouse flew onto a spruce limb, where it sat motionless to watch me.

A species of grouse lives at every level in the Rockies, and an old mountain hand can tell his elevation by them. Sage hens nest in the low hills and sage plains, as their name implies. During mating season the male puts on a remarkable courtship display. He performs an Indian dance in which he hops about and inflates two yellow air sacs on his neck. By expelling the air he makes a loud popping sound, like a balloon bursting. Ruffed grouse inhabit the foothills among the Gambel oaks; the drumming of their wings resembles an outboard motor starting. Higher up, in the conifer forests, live the spruce grouse, with dappled black and brown plumage. Blue grouse are a subalpine species; the presence of this one on the tree limb meant that I was getting close to timberline, and to Balsam Lake.

Above the tree line there lives another grouse, the white-tailed ptarmigan, perhaps the most interesting of all. It makes a gentle clucking sound, like a barnyard hen, and changes its feathers to fit the season. Its summer plumage is brown and white, enabling it to blend with the rocks and tundra; in winter the ptarmigan is pure white and almost indistinguishable against the snow. So effective is the bird's disguise that one of America's first ptarmigan *aficionados,* the 19th Century naturalist Enos A. Mills, had a terrible time attempting to study them. While hiking across the Colorado tundra one day, he spotted two ptarmigan and then quickly lost them. "Not locating either of the birds, I returned to the spot where one had been," he reported. "I had about decided to give up the search, when one of them commenced to peck my shoe. I was standing so close that I was actually touching her with my toe."

New varieties of bright-colored flowers also signaled the rise in altitude: I paused by a stand of monkshood, a biological cousin of the garden delphinium, with vivid blue flowers shaped like a monk's cowl. They nestled like a tiny blue lake in the center of a small glade. Then I took another step, and across a gentle rise, through the spruce branches, was a real lake.

The color of Balsam Lake astounds the senses. So intense is its blue that ink, not water, might be the fluid that fills it. An iridescence plays across its surface like the sheen of a peacock's tail, blending and separating, giving back the viridian reflections of the spruce trees, the green-gold of the sedges, the yellow bottom sand refracted and transformed, a delicate rose and violet from the surrounding cliffs. Some rivers in India mingle as many different colors, downstream from where the women dye their saris.

Balsam Lake, 11,452 feet high at the edge of the timberline, was scooped out by the passage and retreat of high mountain glaciers. The lake's extraordinary color comes from the fact that it is particularly deep. All the sediment sinks far to the bottom, so that the surface water, for the first 60 feet or so, is as clear and reflective as a prism.

The Colorado Mountain Club had pitched its tents on a low bluff overlooking the lake in a glade surrounded by Engelmann spruces. Mountains closed around the campsite like a castle wall. Three of the Grenadier summits, called the Trinity Peaks, jutted beyond the timber to the north, blocking out the view of three others: Vestal, Arrow and Storm King. To the south, on the opposite shore of the lake, rose the escarpments of another mountain cluster, the San Juan Needles. I set up my tent and walked over to the cookfire where some CMC members were plotting the next day's climb.

"We could make all three Trinities, with no sweat," said a trim fortyish gentleman in workshirt and knickers, with the charcoal smudge of a three-day beard.

"Vestal, too, if we started early enough," added a younger man, clean-shaven and in bluejeans, with shoulder-length hair.

"That's a long day," said a third. He was clad only in shorts and T-shirt, though the air was crisp. His flourishing blond whiskers were clearly a long-term project. "We could get a group up Vestal day after tomorrow." Two young lady members murmured assent.

"How about Arrow again?" the long-haired member said. The CMC advance party had attempted Arrow the day before, and the climb had not been pleasant. Loose rock had made the footing treacherous and there had been an overhanging ledge that could not have been crossed except at great risk. The climbers had been forced back before they could reach the summit.

"Sure. Anyone who wants to can try Arrow again," said the blond-whiskered man, who seemed to be the outing's leader. "Then the newcomers could hike up the valley to Storm King. It's a good mountain to warm up on—not a particularly hard climb, and you get the best view of the Continental Divide." This last bit of intelligence was obviously directed at me.

"Sounds good," one of the girls put in. "Storm King is supposed to be a class-three walk-up."

Climbers grade mountains on a numerical scale according to difficulty, something like the Richter scale used to gauge the ferocity of earthquakes. Classes one and two are so easy that experienced mountaineers do not think of them as proper climbs. Class-three mountains demand a certain agility, endurance and some occasional caution. But serious climbers do not consider them particularly worrisome. On a class-four slope you can hurt yourself; well-organized mountaineers such as those of the CMC insist that class-four climbers tie themselves together with a rope. Class five requires every conceivable safety measure, in the form of ropes, spikes, pitons and other ironmongery.

The sense of danger is an important ingredient in the urge to climb, and every now and again a disaster occurs that reinforces it. The month before our Grenadier outing the CMC had undertaken a climb in the Sangre de Cristos, a mountain range about 100 miles east of the San Juans. A novice woman climber had untied herself from the rope against the advice of her companions. She hunkered down on a grassy ledge to rest a moment, and suddenly began to slide. She slid over the edge and into space. Her body plummeted 240 feet down a precipice and onto a pile of rock. One of the CMC members now standing around the cookfire had been there, but he did not like to talk about it. "It was hard carrying down what was left," he said. "We had to put her arms and legs back together again."

The sun had dropped behind the spruces and the air turned chilly. People began pulling on sweaters and parkas or shivering bravely. "Those Himalayan climbers must be downright fanatical," someone said, "freezing off their fingers and toes and staggering about with no air to breathe."

"I'll bet we seem a bit cracked to some people," another man said, edging closer to the fire. Assenting grunts, and a lapse into silence.

"Okay, breakfast at 7, with first call at 6:45," the blond-whiskered man said. "Arrow climbers should be on the road by 8:15. Those for Storm King can leave a bit later, like 9 o'clock."

The morning of the climb up Storm King broke crisp and clear. A thin dusting of frost coated the grasses and wild flowers around the tents, and in the short walk to the cookfire a man's boots became stained with moisture. It is a rule of thumb in the mountains that air temperature drops 3° to 5° for every 1,000-foot rise in elevation. The previous day at Durango, as we boarded the train to Silverton, the morning sun at 6,505 feet was heating up to a midsummer scorcher. Today, almost a mile higher in the Grenadiers, we huddled in our parkas drinking coffee to keep warm.

The sun crept down the flanks of the Needles across the lake, lit the tops of the spruces and thawed the grasses in the campsite. The eight climbers who made up the Storm King expedition began stuffing small day packs full of gear: sweaters, cameras, binoculars, topographical maps, canteens, sandwich lunches and finally our parkas. One climber—the gentleman in knickers—strapped on an ice ax and a 150-foot coil of rope. Though most snowbanks had melted from the San Juans by now, an ice ax could be handy for jamming into steep slopes to stop a slip. The rope was also a precautionary measure, though in theory it would not be needed on a third-class climb such as Storm King.

Our approach would take us east past Balsam Lake, above timber through a boggy meadow, and then up a high valley to a saddle that runs between Storm King and another mountain: Peak Eight of the Needles, directly to the south. The hike would be short—just two miles—but it would require a bit of effort. We would have to gain about 1,300 feet in elevation from the lake to the saddle. Storm King rises above the saddle another 922 feet. We would pick out the route to the summit when we got there.

Thin wisps of cloud were gathering at the edges of the sky as we set out toward the meadow. Mountains this high make their own weather. On a summer morning such as this one the sun heats the lush mountain valleys, warming the air and allowing it to pick up moisture. The hot air rises, blowing up the valleys toward the peaks, carrying the moisture with it. We could feel the updraft gusting gently against our backs. When the moisture-laden air reaches the heights, it cools and the water vapor condenses into clouds. By midafternoon the sky may boil with ominous gray thunderheads, and rain or hail will start.

Nestled beneath the Trinity Peaks of the San Juan Mountains, Balsam Lake appears green from the reflection of Engelmann-spruce trees.

It is not a good idea to get caught on a mountaintop in a thunderstorm. Rain makes footholds slippery and treacherous. A storm's gusty winds do not help a climber negotiate a rock face. Lightning in the high mountains is more awesome and more highly charged than anywhere else on earth. Consider the experience of Franklin Rhoda, one of the first San Juan climbers. Rhoda was a member of a United States government survey team, led by Ferdinand Hayden, that in 1871 made some of the first official maps of the Rocky Mountains. Here is what happened when he set his transit on top of 14,001-foot Sunshine Peak, just north of the Grenadiers. Clearly the sun was not shining on Rhoda and his survey team companion, A. D. Wilson.

"We had scarcely got started to work when we both began to feel a peculiar tickling sensation along the roots of our hair, just at the edge of our hats, caused by electricity in the air. By holding up our hands above our heads a tickling sound was produced . . . like that produced by the frying of bacon. . . . The instrument on the tripod began to click like a telegraph machine when it is made to work rapidly; at the same time we noticed that the pencils in our fingers made a similar but finer sound whenever we let them lie back so as to touch the flesh of the hand between thumb and forefinger. . . . Wilson was driven from his instrument, and we both crouched down among the rocks to await the relief to be given by the next stroke [of lightning], which, for aught we knew, might strike the instrument which now stood alone on the summit. At this time it was producing a terrible humming, which, with the noises emitted by the thousands of angular blocks of stone, and the sounds produced by our hair, made such a din we could scarcely think."

Wilson thought enough of his transit to scurry back to the summit for it, and was rewarded by "a strong electric shock, accompanied by a pain as if a sharp-pointed instrument had pierced his shoulder," which luckily did not cause any permanent damage. There are techniques for riding out mountain storms, but running along an exposed ridge with a surveying transit is not one of them. Nor is it sensible to look for shelter under an overhanging crag. Lightning may jump between two protruding rocks or from an overhang to the ground, the way a spark leaps between the two terminals of an electric arc. The best plan is to find a slope where the ground is gently concave and crouch, making yourself as small—and keeping as dry—as possible. The rain will beat down on you, but not the lightning.

No thunderheads had gathered as we walked out across the meadow. The sun poured down on the bright clumps of wild flowers and sedges that covered the marshy earth. Stunted willows caught at our legs as we forced our way through them. Occasionally a boot would break through the grass matting into a bog.

Mountain meadows often start as lakes that gradually fill up with decayed vegetable matter and soil. The process is called eutrophication, and it takes thousands of years—except in civilized lands where people flush their phosphate detergents into the sewer system. Phosphates are a powerful fertilizer and aphrodisiac for fresh-water algae, sending it on a reproductive rampage that can choke up a lake centuries before its time. But no phosphates had tainted the pleasant acreage we were moving across. This was simply a natural interval in the evolution between bog and meadow.

The meadow ended in an immense rockslide. Boulders that ranged in size from beer kegs to packing crates to Volkswagen cars were tumbled one on top of another all the way to the lip of the alpine valley, about 400 feet up. Such a formation is called a talus slope and it results from a type of erosion caused by the cyclical baking and freezing of mountain bedrock. Even when the air is cold, radiation from the sun at high altitudes can heat rock to an intense degree, making it expand. When the sun sets, the temperature usually dips below freezing, and the rock contracts. These alternating forces open tiny fractures so that water can seep in, expanding as it freezes and prying loose the rock. Thus mountaintops are subjected to a gigantic thermal engine, run by the weather, that breaks them apart as effectively as blasting powder.

The boulders on a talus slope are sharp and faceted, like the debris in a stone quarry. Only the hardiest plants can survive in them—lichens that dapple their surfaces like daubs of orange and green paint, a tiny yellow stonecrop or a clump of blue-and-white columbine where a cupful of soil had lodged in a crevice. But you do not look at plants on talus slopes. As our group of climbers toiled single file toward the lip of the alpine valley, our eyes were on our feet and where to place them amid the rubble.

Not so when we reached the valley itself. The land above the tree line is strange and remote, almost beyond description. The stupendous crags of the Grenadiers loomed steeper than ever above the valley floor. Balsam Lake, a diminished blue glint among the trees, seemed as far below us as if seen through an airplane window. I felt as removed from civilization as an explorer on the surface of an asteroid spinning in space between Mars and Jupiter. Other climbers might have come along this way earlier in the summer. If so, they had left no sign.

The valley floor was patterned in weird shapes—domelike hummocks, round depressions that held mounds of loose stone, grassy sections enclosed within shallow stone moats. A pinkish bedrock, buffed to a mirror finish by the glacier that had scoured through the valley, protruded from the thin soil like the foundation stones of a ruined building. The only visible life was a thin tissue of sedges, herbs and dwarf willows no higher than a boot sole. This was tundra. It is found in some of the most remote places in the world—the Far North near the Arctic Circle, and in mountains above timberline. It is present over long stretches of the Great Divide.

Spring and summer on the heights of the divide bring forth a variety of wild flowers that flourish in spite of limited moisture, short growing seasons and the constant buffeting of strong winds. Each flower shown here has its way of coping with the elements. The yellow alpine sunflower waits several years before blooming, then blossoms once and dies; the arctic gentian, availing itself of August warmth, is a late bloomer in timberline meadows; rosecrown thrives among rocks in very little soil. The Colorado blue columbine—the state flower—prospers above and below timberline, growing up to 24 inches tall.

Tundra is worth stopping for. It occurs under the most rigorous conditions of climate. Half the year it is covered with snow, except where winter winds sweep it bare. About seven miles to the east of the Grenadiers, on a ridge flanking the Continental Divide, the University of Colorado keeps a cluster of all-weather tents where scientists of the Institute of Arctic and Alpine Research measure things like frost depth, plant growth and climate in the tundra. The year-round average temperature, they estimate, is just under 30°F. and the number of frost-free days can be as few as 60. Along some parts of the divide, at certain soil depths, the frost never leaves, even in summer. It seems a miracle that tundra plants can grow at all.

Yet they do, and in a profusion and variety that astonishes. Wild flowers glitter across the valley floor like cut gems: the ruby of king's-crown, sapphire of gentian, topaz of stonecrop and avens. Some plants stand in thick leafy clumps, such as the Parry's primrose that thrives where the meltwater trickles down from a snowbank. Ebony sedge, which resembles grass but belongs to an entirely different family, thrusts up miniature black pompons on the drier, more exposed flats. From a barren rock face blooms a preposterously showy flower called a polemonium, which puts out what can only be described as a pandemonium of azure petals.

Each plant has found a way to adapt itself to a particular nook of the tundra environment. Some do it by fast living. The alpine saxifrage takes only five days to shoot out new leaves, send up a flower and disperse its pollen. Others are rock-bound conservatives. The alpine phlox hugs the earth in a tight round cluster of minute leaves and small white flowers—an arrangement that reminds one of a green pincushion with white embroidery. The compact leaf structure holds in the heat, so that

ALPINE SUNFLOWER

ARCTIC GENTIAN

ROSECROWN

COLORADO BLUE COLUMBINE

the temperature inside the cushion may be 20° higher than the outside air. The phlox also conserves its strength, growing only a leaf or two each year. A plant the size of a fist may be 150 years old.

Even the colors of these alpine flowers serve a utilitarian purpose: they attract the bees that carry pollen from one plant to another. And odd as it may seem in a world so spartan, there are bees on the tundra. As I sat by a rock to rest, midway up the valley to Storm King, a bumblebee sat down beside me on the lip of an arctic gentian. These late-summer flowers, inconspicuous white blossoms shaped like ragged trumpets, were just opening up and the bee seemed to be specializing in them. He blundered from one to the other as busily as any lowland bumblebee in someone's garden. His presence among these bleak and tumbled rocks was positively surreal.

For the bee it was all in a summer's work. Each year swarms of bugs, beetles, bees and butterflies ride up from the prairies on the wind. Recently at a station of the Institute of Arctic and Alpine Research on the Continental Divide near Boulder, Colorado, the resident ornithologist was confronted with a puzzling discrepancy in his study of the ptarmigan. The fat and nerve tissue of ptarmigan chicks, he found, contained a small percentage of DDT. Far less insecticide was found in older birds, however. The ornithologist knew that while adult ptarmigans are vegetarians—they eat the leaves of alpine willows, for example—newborn chicks often feast on larvae and insects that are found in snowbanks. But no one has ever sprayed snowbanks or tundra with DDT. The explanation, provided by an entomologist, was absurdly simple. The baby chicks had been eating insects blown up from the plains, where an entire generation of farmers had dusted crops with DDT.

Mountain climbers profess great fondness for bees and flowers, but they seldom stop to look at them. If you want to see how many impossible peaks you can conquer in a day's outing, there is simply no time. People who dawdle are called mushroom pickers. And though the fraternity of climbers is shot full of mushroom lovers, both gourmets and botanists, there is the implication that the act of mushroom picking is not quite suitable for a red-blooded American mountain man. I started up the valley, moving quickly to catch the rest of the climbers, who were already nearing the saddle at the foot of Storm King.

The walk to the saddle took perhaps half an hour, and by the time I had reached it the bright morning weather had taken a turn for the worse. The wisps of cloud we had noticed earlier had now congealed into dense gray thunderheads. An icy wind seemed to blow from all directions at once, caroming off the peaks and down onto the saddle. Three CMC members sat bundled in parkas among the rocks, eating chocolate bars. They had decided to wait here while the five other members of the party, myself included, climbed to the summit of Storm King without them. And no wonder. The mountain rises above the saddle almost 1,000 feet, with walls of eroded bedrock and loose talus that appear, to a person standing at their foot, to go straight up.

Steepness obviously did not deter the rest of the assault team. They were already climbing, and had advanced about 200 feet up the saddle and onto the mountain face. I have always wondered why, when tramping through wilderness, people invariably mimic the footsteps of the individual directly in front of them. Even when there is no trail, and no need for one, hikers march across the landscape single file, as though coupled together like railroad cars. Yet men search out wilderness to free themselves from other men. What ancient protective urge is buried in our genes that causes us, in a strange land, to follow each other like a string of sheep? Suddenly, watching the other climbers disappear up the precipice, I knew. It is the panic of being left behind.

I darted up the loose rock of the saddle as quickly as my legs would allow. The route followed the crest of the saddle onto a steep ridge of bare rock, deeply eroded into crevices and knobs. The ridge rose nearly to the summit, and indeed seemed to support it, like a buttress on a medieval church. To the left of the ridge the mountain face disintegrated into a precipitous chute of loose rubble, with rock chunks that varied in size from angular slivers resembling potsherds to boulders that must have weighed several tons. The rubble was both steep and unstable, needing only a gentle nudge to make it slide down the mountain a bit. Nonetheless, the two lead climbers had left the ridge about halfway up and were advancing cautiously along the rubble. As they walked, each footstep set free a small avalanche of loose rock.

The scramble to catch up had left me prematurely winded. None of the climb so far had been particularly complicated, but it was exhausting uphill work. I had struggled about 400 feet above the saddle and could see over some of the immediate peaks to crests and pinnacles in the distance. The three people on the saddle had shrunk to toy figures, with only the bright colors of their parkas to distinguish them. The other climbers had vanished behind an outcrop that separated me from the rockfall. All except one—the gentleman in knickers, who was moving slowly and deliberately about 30 feet above me. Perhaps the added seven pounds of nylon climbing rope hooked in a coil to his rucksack was cutting his speed. In any case, it was good to find company, particularly since the ridge ahead was beginning to steepen.

And beginning to fall apart. Nothing I touched seemed to remain in place. When I grabbed for a purchase, the mountain would come off in my hand. A rock falling off a mountain face gathers speed as it goes, bounding and careening off other rocks until it disappears from sight. Each hit makes a sharp report, like a rifle shot, that echoes from crag to crag. Two or three rocks falling at once explode like a small war.

I let my knickered companion explore the route. Somehow or other we had worked off the ridge to the margin of the rubble. We would find a launching point in fairly stable rock and then dash up the loose stuff to an outcrop that looked as though it might hold us. Each scramble sent down a cascade of stone missiles that would certainly have knocked over anyone directly below. To my eye it seemed that any falling body, inanimate or human, would not stop until it reached the saddle. The mind in such circumstances tends to exaggerate. But knowing the height of Storm King and our position on it, I judged that the distance to the saddle was about 700 feet—straight down. The Golden Gate Bridge in San Francisco, from the top of the towers to the water, is a little more than 700 feet.

Orange, yellow and black lichens —hardiest of the plants that brave the mountains' high altitudes—encrust a boulder on the alpine tundra, terrain too harsh for trees. Above and below the rock are two other tundra natives, alpine clover (top) and alpine sandwort.

There is no word in a mountain climber's vocabulary for acrophobia. The closest approximation is the term "exposure." A climber may say, "that face has a certain degree of exposure." He means that if you slip you travel a long way before you hit bottom. My legs had begun to suffer from exposure. After each step they vibrated like jackhammers. My sense of equilibrium was failing. My day pack, though it weighed only 15 pounds, seemed determined to pull me backwards off the mountain. The slope below me angled down for about 100 feet and then dropped out of sight. I could make out the silhouettes of the rocks on the lip, and beyond them, space.

I knew I could get myself down the mountain from where I sat. I also knew, with the faith some men have in judgment day, that I could never reach the next outcrop. A cube of quartzite about the size of an egg crate crashed down the scree and over the edge. Howitzer fire echoed back up. An odd temptation presented itself. By placing the palms of my hands against the mountainside and pushing, I would be able to catapult into space. For one glorious moment my body would ride the thermal currents as gracefully as any eagle. Then downslope, in a cavity I could not see, a crumpled green parka would ooze liquid like a blemished plum, and a new kind of sticky red lichen would blossom across the rocks.

"Well, we've near got her licked!" This cheerful shout came from my climbing companion. "Couple of hundred feet and you'll be able to spit to the top."

There was no way to answer. I could neither open my mouth nor close it. Somehow my tongue had grown a jacket of flannel that absorbed every bit of saliva. In any case, you do not contradict a madman. Did my knickered friend not understand how precariously we clung to survival? Clearly not, for he bounded along as blithely as a parson at a barn dance. Yet some tremor of alarm must have projected itself, for a minute later he was poised beside me delivering a cram course in basic mountaineering. "You tie the end of this rope around your waist with a double bowline—like this. Then you wait here while I take this other end up the mountain. When I holler, you just follow up the rope."

What if I stumbled and carried him with me? "Can't happen," I was told. "I could anchor 1,000 pounds with this rope. I weigh 193, and my eight-year-old son can hold me."

I did not ponder the strange physics of roping up. I simply followed, when the holler came, clutching, sliding, grappling, leaping, clawing in an adrenalin frenzy up the precipice. When I reached a solid perch near my friend, we repeated the maneuver. I braced myself, he climbed, and on signal I followed. A third time, and we were crouched side by side on the summit.

The top of Storm King is the thin end of a wedge-shaped ridge, narrowing in places to about two feet, with an abrupt 1,300-foot plunge on the north side. To reach it we had ascended 948 feet from the saddle. We now stood 2,307 feet above Balsam Lake, 7,237 feet above Durango and 13,742 feet higher than the surface of the ocean.

Altitude does extraordinary things to people. I have a friend from college days, a brawny six-footer who, at 11,000 feet in the Alps and for no discernible reason, collapsed among the edelweiss and sobbed like an infant. Other people have been known to grow dizzy, get migraines, black out or regurgitate their breakfast. Sometimes the response is uncontrollable euphoria. In 1868 the Lieutenant Governor of Illinois, William Bross, accompanied by his friend Schuyler Colfax, Speaker of the House of Representatives, and other assorted ladies and gentlemen, felt so lightheaded on attaining the 14,286-foot summit of Mt. Lincoln in central Colorado that he led the group in an impassioned rendition of the doxology: "Praise God from Whom all blessings flow" resounded across the tundra to an audience of hawks and eagles.

I had started shaking—not from fear now, but from a fit of giggles. It was as though someone had given me a whiff of an intoxicating drug. Relief, joy, love, hysteria, omnipotence, awe—a Mixmaster of emotions whizzed inside my head. Not many people, considering the multitudes that have walked the earth since men descended from the trees and took to standing on their two hind legs, have seen the world from this particular spot. Several thousand perhaps, give or take a few hundred.

It was an awesome, exhilarating panorama. Wave upon wave of mountain ranges rolled into the distance, green and tan close by, receding to violet and gray, and then shadowy blue washes at the foot of the sky. Gray cloud banks fumed above the peaks, casting some into shadow, opening patches of blue above others, so that the sun would spot them like a klieg light. We counted three separate thunderstorms all going at once, the curtains of rain obscuring the ridges behind them like theatrical scrims.

The Great Divide lay below us to the east, an undulating saddle of green velvet. It was the same tundra of sedge and wild flowers we had walked across in the valley on the way to Storm King. The gap between us was only three miles wide, but the land was so rugged that a full day's hike could not have taken us there. Across the saddle lay the headwaters of the Rio Grande, hidden among the near peaks. Beyond them the divide curved eastward along an extended cliff edge that was bisected by a rectangular cut. The cut resembled a doorway with the lintel removed. This was a major divide landmark, the Spanish Window, so perfectly symmetrical that it might have been shaped by a stonemason with T square and plumb line. Beyond the window the divide rose to the 13,838-foot summit of the Rio Grande Pyramid and then disappeared, dropping southward along the Rio Grande Valley.

We could not stay long. Flurries of snow, soft compacted pellets the size of rice grains, blew across Storm King. Thunder crackled in the distance, and unless we wanted to repeat the electrifying experience of Franklin Rhoda on his survey, we would have to get off the mountain. Yet my fear had utterly evaporated. The slope, so awesome on the way up, had somehow lost half its pitch. I bounded down it as confidently as a mountain goat. No matter that each leap would start a small rockslide; we rode the rockslides the way surfers ride waves.

My elation held as I reached the saddle. It buoyed me down the valley and across the tundra toward Balsam Lake. It persisted through the waning afternoon as I sat by the cookfire with a cup of tea. It continued as one of the other climbers, a doctor, gave me a medical explanation.

The euphoria of high places has a name, it seems—hypoxia. The word can be translated as oxygen shortage. Hypoxia results from the fact that air at high altitudes is much thinner than the air people normally breathe. At sea level, where air pressure measures 15 pounds per square inch, each cubic foot of it weighs 1.2 ounces. If you climb to 14,000 feet its weight drops to 0.8 ounces per cubic foot, about three fifths as much. Each breath you take, then, gives you only two thirds your usual amount of oxygen. Your body tries to compensate: your lungs work harder, your pulse rate soars from 72 beats per minute to more than 100 and your bone marrow starts manufacturing more red blood corpuscles. But until you become fully acclimated, a process requiring about three months for most people, your body is oxygen-starved. Your stomach cannot digest efficiently—thus the nausea some people experience—and your brain cells click along at half speed. Dizziness and headaches occur—or an intoxication as beguiling as the high from alcohol or drugs.

But there had to be another reason that had nothing to do with blood corpuscles and brain cells. The Indians who lived in the shadow of the Rocky Mountains, and in whose vocabulary medicine and magic are virtually the same, held mountains in great reverence. It was the custom for young warriors, at a certain time in their lives, to set out from the tribe and travel alone into the high places. There they would stay, fasting and praying, until whatever spirit inhabited the area granted them a vision. It might be a dream of a mountain lion or a buffalo or an eagle flying into the sun. Such visions were powerful medicine, and the brave would return to the tribe a stronger and illuminated man. Earlier that day, more than two and a half miles above sea level among the thunderheads on the hatchet-edged summit of Storm King, I had felt perhaps an echo of the same thing.

An odd thought, brought on by an odd necromancy of altitude and scenery. I sat by the cookfire with my tea and watched the clouds roll back from the sky and the evening darken. The oblique rays of the falling sun glowed on the peaks across Balsam Lake like a benediction.

A Flourish of Autumn Aspen

PHOTOGRAPHS BY BOB WATERMAN

One of the most stunning sights in Great Divide country is the fanfare of orange, chrome yellow and crimson that is emblazoned across the landscape every autumn by legions of quaking aspen trees.

The aspen is a hallmark of the Western mountains. Known also as trembling or golden aspen, the tree looks a bit like birch, with its white bark and slender straight trunk, but is no relation. It is more distinguishable by its leaves, which are green on top and silvery underneath. They are balanced on long flexible stems and quiver at the least provocation, so that in even the faintest breeze the whole tree appears to be magically shimmering.

In summer the aspens' pale green foliage stands out in delicate counterpoint to the somber conifers that clothe the hillsides. In autumn the aspen, like the maple, is decked in every shade from deep red to golden yellow, with yellow predominating. The varying colors are attributed to diverse local weather conditions, to differences in soil composition and to genetic variations between one clump of trees and the next. In fact, the divergences in the trees' genetic characteristics are analogous to the differences of skin pigmentation or hair color among families belonging to the same race.

In the spring aspens produce great quantities of seed, each with a tiny tuft of fluff that bears it on the wind for miles. Only a small fraction of them ever take root, but even so, aspens reproduce prolifically: each tree's root system has hundreds of dormant buds that are quick to sprout when encouraged by the sun's light and warmth.

In this way aspens can multiply at an astonishing rate: as many as 200,000 new shoots have been counted in one acre of open ground. Furthermore the saplings can shoot up more than four feet in their first three years—in 30 years they may reach 90 feet. Because of this rapid rate of growth, aspens serve as the principal agent in the reforestation of burned areas. As they grow they provide a nursery for the protection of slow-growing trees; evergreens and some hardy species of oaks take root under the aspens' canopy and thrive on the rich humus provided by the aspens' fallen leaves. In about 50 years the younger trees begin to crowd and then overshadow the aspens, which soon die and yield their ephemeral beauty to the vigor of the evergreen forest reborn.

Flamboyant against the mist of the San Juan Mountains, a stand of quaking aspens brightens a cloudy valley on the western flank of the divide.

The luminous yellow of the foliage in the picture below is the aspen's most characteristic fall color and has earned it the name of golden aspen.

The reddish hues of these aspens will appear in successor trees if the parent trees are not cross-pollinated with others having different-colored leaves.

Reaching the end of their autumn cycle, the slender trees of a young aspen grove cluster in bare symmetry, already denuded of their foliage. The dead leaves, spangled by raindrops (above), form a thick carpet that will enrich the soil and prepare it for the growth of a future stand of evergreen trees.

4/ Haven of the Wild Horse

South Pass...is the one watered place in the entire two thousand miles of Rocky Mountains where wagons and animals and people can cross the chain without climbing anything. MARSHALL SPRAGUE/ *THE GREAT GATES*

There is no good reason, I suppose, for trying to disguise the whereabouts of the Great Divide Basin. You just cannot hide an expanse of 2.25 million acres, no matter how remote those acres may be. I might tell you, by way of discouragement, that the region is unspeakably desolate, which is true. That no trees grow there, also true. That it contains very little water, lots of dust and a wind that never seems to stop. It is not a place most people would choose for an easy Sunday outing. This is fine by me. When crowds of visitors start trekking through the Great Divide Basin—which, if you must know, is where the Continental Divide cuts through southwestern Wyoming—the land will change. The silences will diminish, the loneliness will dissipate, and the vast empty spaces will lose their special aura of freedom and discovery.

No place I have ever been seems so totally removed from the ways of men. Here is the land of the American West as it must have appeared to the first explorers and pioneers. Bleak sage wastes and alkali flats, dry creeks and sand dunes, buttes and badlands, with a fringe of rolling hills as thinly vegetated as a bald man's skull. Most of it is public land, belonging to no one really, unless you count the Bureau of Land Management in Washington, D.C., which administers it with an exceedingly light touch. There is no National Park Service to issue a campfire permit, no forest ranger to name the spring flowers for you, no professional outfitter to trot you out there on horseback. Its true possessors are the natural ones: antelope, coyotes and wild horses.

Such was not always the case. In the middle decades of the last century this barren and melancholy region swarmed with travelers heading west across the Continental Divide. The procession threaded along the uplands beyond the basin's northern rim, an interminable caravan of packtrains, stagecoaches, Pony Express riders, prairie schooners, and even pedestrians who had walked all the way from Missouri trundling their possessions in handcarts and wheelbarrows. For just north of the basin, at a place called South Pass, lay the main emigration route to Oregon and California. Here the Oregon Trail crossed the Great Divide.

Other pathways led over the mountains, to be sure. Indians had found passages across the Rockies before the first white men ever came. And the trail to Santa Fe and beyond, which crossed the divide in New Mexico, had been an important trade artery ever since 1823. But along most of its length the divide, for anyone trying to cross it in a wagon with his goods and chattel, was impassable.

At this one place, however, the barrier breaks down. Just south of the Wind River Range in Wyoming the mountains disappear and the arid high plains roll through the gap toward Utah. Here the divide leaves the domes and pinnacles of the Wind Rivers, drops into the foothills and meanders across the flats. For a north-south distance of about 25 miles, the ridge of the continent rises so imperceptibly that a traveler cannot tell when he has reached its summit. This gap is South Pass.

Below the pass the divide rises again to a pair of rocky eminences, called the Oregon Buttes, that mark the place where the Continental Divide meets the northern lip of the Great Divide Basin. Here the divide does a curious thing: it splits in two. The branches separate, one continuing south and the other veering east, each going its own way until they rejoin 90 miles or so to the southeast near the Colorado border. The two arms encircle the basin like the lip of a saucer, and in fact define the basin's perimeter. It is as though a warp had occurred in space, opening a cavity in the infinitely thin line of the divide and allowing the basin to nestle inside. The divide still parts the waters of the continent, but in an odd way. All the rain falling outside the basin flows toward either the Atlantic or the Pacific Ocean, as it normally would. But the bare six to eight inches of rain that generally falls inside is caught: it just stays there until it evaporates in the basin's desert heat. No wagon boss worth his salt ever took a train across this parched region, but held instead to the route through the pass.

The trail through South Pass took some finding, however. Meriweth-

In single file, four wild horses cross a plain in the Great Divide Basin. Descended from once-domesticated stock and now protected by federal law, the mustangs will briefly tolerate inspection at close range (above), then gallop off to resume their roamings.

er Lewis and William Clark, the first United States citizens over the divide, crossed the barrier far to the north, in what is now Montana. The two men were young and energetic, with all the optimism of early manhood, and the prospect of bushwhacking through a few mountains did not bother them at all. In fact, they planned to canoe across.

The idea did not seem so fantastic at the time. No one—except the Indians—yet knew how rugged the Rockies could be. The most promising route into the West, the one taken by Indian traders and French beaver trappers, went up the Missouri River. Most of these early *voyageurs* used wooden rowboats called pirogues. Lewis and Clark planned to start out more or less the same way, with a 32-man entourage of frontiersmen, soldiers and French boatmen. When the Missouri became too shallow for their pirogues, they intended to try out a contraption invented by Lewis—a collapsible iron-framed canoe that could be dismantled for a brief portage from the Missouri headwaters to the first Pacific-bound stream. Once over the hump they expected to coast down the Columbia River to the ocean. The scheme was the last echo of an ancient refrain in American exploration—the search for a Northwest Passage, by water, across the continent.

The plan worked well enough until the expedition reached the mountains. There the Missouri developed so many shoals that the pirogues had to be abandoned prematurely. The collapsible iron canoe proved totally useless; it leaked so dreadfully it would not even float. "The boat in every other rispect," Captain Lewis noted stubbornly in his journal, "completely answers my most sanguine expectation."

Luckily the two explorers had picked up a guide, a Shoshone girl named Sacajawea, who helped them find an Indian trail through the high country. On August 12, 1805, in the Bitterroot Mountains that today mark the border between Montana and Idaho, the expedition arrived at the Great Divide. "The road took us to the most distant fountain of the waters of the Mighty Missouri, in search of which we had spent so many toilsome days and wristless nights," wrote Lewis, who was no great hand at dictionary spelling.

"After refreshing ourselves we proceeded on to the top of the dividing ridge," Lewis continued, "from which I discovered immence ranges of high mountains still to the West of us with their tops partially covered with snow." Clearly it was not boating country. But the explorers tried anyway, with animal-hide canoes. Clark floundered down a creek on the west slope, where he "found the river shoally, rapid, shallow and extremely difficult. The men in the water almost all day. They are getting weak soar and much fortiiegued."

There had to be a better passage through the mountains. And so there was, though the man who discovered it came through by the back door, from west to east, in snow and biting cold—an experience so thoroughly miserable that for many years no one repeated it. Robert Stuart, a volatile Scot of 27, was a member of a trading post established via sailing ship on the Pacific Coast by John Jacob Astor, the fur magnate. In June of 1812 Stuart set out from Astor's post at the mouth of the Columbia River with six men, carrying dispatches. At first the trip went smoothly enough, until Stuart met an Indian who promised to show him a short cut over the divide—for a fee. Stuart made the mistake of paying in advance. The Indian quickly made off with his loot, leaving Stuart to find the way by himself. Soon after, Crow Indians stole all the horses. Game became scarce. The men, hungry and exhausted, started bickering. The food ran out entirely and someone proposed that the company draw lots, the loser to be shot and cut up into steaks. "I shuddered at the idea," wrote Stuart, a staunch Presbyterian even when starving. "I snatched up my Rifle, cocked and leveled it at him . . . he fell instantly upon his knees and asked the whole party's pardon."

The next day brought relief: the expedition shot a buffalo, "and so ravenous were our appetites that we ate part of the animal raw." By this time it was autumn and the snow started falling. Then, on October 22, Stuart came to a stream that flowed east. He had crossed South Pass, the first white man in history to do so. About 100 miles east of the pass the party dug in for winter, and did not reach St. Louis with the dispatch case until the following April.

Nobody paid much attention to Stuart's discovery. By the time he arrived at St. Louis the United States had become too busy fighting a war with England to think about new routes across the Rockies. Besides, Stuart's trip had hardly been a model of ease and convenience. The South Pass region was not touched again for more than a decade.

In 1824 a Bible-toting mountain man named Jedediah Smith stumbled upon the pass while heading west with an expedition of trappers. Smith's worst misfortune was an encounter with a grizzly bear, which clawed off part of his scalp. But Smith was rugged; he had himself sewn up and kept on trekking.

Soon South Pass became the standard thoroughfare to the rich beaver streams west of the divide. Each summer the fur companies would load supplies for their trappers onto packtrains and lead them across the prairies and over the pass. Each fall the packtrains would return with many thousands of dollars in beaver pelts. Eventually the first wagons heaved across, driven by pioneers bound for California and Oregon. They came in small tentative groups at first, a mere trickle, then in streams and finally in floods—300,000 from 1843 to 1869.

The movement through South Pass was part of the mightiest redistribution of humanity in the young nation's history, a mass exodus across the Great Plains to the lands beyond the mountains. The travelers were called emigrants because they were actually leaving the country. The main trail began where the United States ended, at Independence, Missouri, which in 1843 was the westernmost town in the Union's westernmost state. It ran more than 2,000 miles out through Indian and buffalo country, along the prairies of the Nebraska Territory and up the high arid plains of Wyoming to the pass. Beyond this lay the long pull through deserts and mountain defiles to California and Oregon, where land was free, where streams were paved with gold and where winter never came—or so the emigrants believed.

To spare their precious draft animals most people trudged the endless miles beside their wagons, the monotony broken only by crisis—an Indian raid, a splintered wagon brace, a draft ox dead of thirst or fatigue. So many cattle perished along the trail that stretches of it could be walked at night simply by following the litter of bleached skeletons glowing in the moonlight.

Big sagebrush, which thrives in arid Western valleys, sometimes grows as tall as 10 feet, like this hardy specimen on the rim of the Great Divide Basin. An important source of food for wildlife, its evergreen leaves help sustain antelope, elk and mule deer through the late winter months.

One traveler at the height of the migration was Mark Twain, who took the trail to Nevada in 1861. Twain went the posh way, by stagecoach. Near South Pass he overhauled an emigrant train of 33 wagons, which he described in *Roughing It,* his book of Western adventures: "Tramping wearily along and driving their herd of loose cows were dozens of coarse-clad and sad-looking men, women, and children, who had walked as they were walking now, day after day for eight lingering weeks, and in that time had compassed the distance our stage had come in *eight days and three hours*—seven hundred and ninety-eight miles! They were dusty and uncombed, hatless, bonnetless and ragged, and they did look so tired!"

Most travelers, by the time they reached South Pass, were too exhausted to take much notice. Few journals mention it and those that do are usually disparaging. "This tract would be good for roads, as it seems absolutely good for nothing else," declared the venerable Horace Greeley, editor of the New York *Tribune* and staunch advocate of going

West, who in 1859 decided to see for himself what the West was really like. "There is scarcely a bushel of soil to each square rod, of course no grass," Greeley complained, and "not a mule feed to each acre."

And so I did not expect very much one July morning not long ago as I drove across the sagebrush toward South Pass in a pickup truck. The day was one of those that the weather bureau calls partly sunny, with thick gray clouds boiling up from the west and fanning out across the sky. A strong wind raised miniature cyclones of dirt and sand, and gusted so viciously you could not look it in the eye. The pickup lurched along a rutted trail in the thin soil of the hills east of the pass.

This was indeed barren land, without a tree in sight. And yet it was beautiful land, the sagebrush gray on the hillsides, the sparse grass a delicate pastel tissue in the bottoms. The country dipped and rose in gentle billows, like the sea after a storm, and as you climbed the higher swells the distances would expand and the sky grow huge. To the north the snowy peaks of the Wind Rivers would lift like a vapor. You could look across such immensities of landscape that space itself became a palpable and vibrant thing.

Yet the land was by no means empty. Antelope browsed on sage in clusters of two or three, the does with their kids beside them. They would raise their heads, stand like statues, then suddenly turn tail, flashing their white rump patches, and disappear. Zoologists do not call them antelope, though everyone else does. Their proper name is pronghorn; real antelope live in Africa and Asia. Pronghorns are related but live only in the American West and Mexico. One of the fastest animals on earth, the pronghorn can reach 60 miles an hour in short sprints.

Other animals scurried about: ground squirrels, gophers, marmots, a coyote. At one point I spotted a golden eagle as he braced against the air currents, searching the ground for his lunch. But most remarkable of all were the horned larks. They swarmed thick as mosquitoes when the truck bounced past them down the trail. There must have been thousands upon thousands nesting in the sagebrush, and they flew with a strange darting motion, folding their wings against their bodies at the end of each stroke. And they do indeed have horns—two clusters of small black feathers jutting from the brow of each bird.

Flowers bloomed, surprisingly profusely for land so bleak. They made an odd mixture: grassland varieties such as daisy and locoweed, a cheerful red foothill species called scarlet trumpet, and such alpine miniatures as stonecrop and cushion pink. Bright blue canopies of larkspur covered the bottom lands where moisture gathered. Larkspurs are wild delphiniums, and for all their beauty contain a poison—delphinine—that is fatal to cattle. There is irony in this. In early spring when other forage is scarce, cattle sometimes eat larkspur and die in great numbers. Later, when grass is plentiful and the cattle have enough to eat, larkspur is no threat; after blooming, it ceases to be toxic.

The information came from my companion Tom Bell, who grew up on a cattle ranch near South Pass and once raised livestock there himself. But he is also a conservationist, which to some veteran cattlemen is a form of professional treachery equivalent to an auto manufacturer's promoting mass transit. In a weekly newspaper he edits, *High Country News,* Tom lambastes cattle and sheep ranchers—perhaps the most powerful group in Wyoming—for poisoning coyotes, shooting eagles and fencing public lands. "There are thousands of miles of illegal fence in Wyoming," he declares. "The fences interrupt the migration routes of wildlife. Antelope get hung up on the wire. The snow catches 'em, and they freeze to death by the thousands."

Tom Bell loves the remote land with a quiet and tenacious passion, and he hopes the land will stay that way. He was thus a bit hesitant about escorting me to South Pass and the Great Divide Basin. He does not want a lot of people touring this country, running their jeeps over the wild flowers and cluttering up the sagebrush with their empty beer cans. But Tom is a realist, and he knows that a sympathetic public will help him in his battle against the fences.

Just before reaching the pass we stopped for lunch on a knoll covered with alpine plants. Tom pointed to an elegantly shaped pink flower without any leaves, growing low to the ground. "There's a bit of history for you," he said. "It's a *Lewisia pygmaea,* named after Meriwether Lewis, who saw it on his trip through the Rockies. The Indians used to eat the root, and I guess Lewis and Clark did too. It's called bitterroot, like the Bitterroot Mountains. Try some." I pulled up the root, a white fiber the size of a toothpick, and chewed into it. Bitter but not unpleasant, like a raw turnip.

We drove on to the pass itself and at first glance it seemed as dreary as the early visitors had indicated. A few sorry clumps of grass poked up through a grayish-white soil like weeds in a parking lot. Yet South Pass is full of echoes and ghosts. A stone marker commemorates the first crossing of the divide by white women. They were Narcissa Whitman and Eliza Spalding, wives of Protestant missionaries, who in 1836 journeyed to Oregon with their husbands to preach the gospel to the Indians. Narcissa happened to be on her honeymoon and pregnant, and it took a fair bit of Christian fortitude to bring her through. "Do not think I regret coming," she confided in her journal. "I am contented and happy notwithstanding I get very hungry & weary. Have six weeks steady journeying before us. Will the Lord give me patience to endure it."

Narcissa's husband, Marcus Whitman, had fortitude enough for everyone. He brought not only the first white woman, but in 1843 helped lead the first big emigrant train across South Pass. He would have been better off staying home. For in 1847 both he and Narcissa were massacred by the very Oregon Indians they were trying to convert.

The evidence of the emigrants' journey through South Pass can still be seen. Two parallel depressions are worn into the soil where the thousands of wagons left their wheel ruts in the trail. The ruts have survived more than a century of punishment by wind and storm.

About 10 miles south of the wagon ruts, at the rim of the Great Divide Basin, rise the Oregon Buttes. Not far from here the ancient trail splits, one pair of ruts trending southwest, the other northwest. At this point the emigrants made a choice—and lived the rest of their lives accordingly. Some took the southerly route toward California; the others held to the north along the Oregon Trail. Now, more than a hundred years later, you cannot stand near this fork without drifting back in time to see in your mind's eye the wagons parting, and to watch the weary backs of the two lines of pioneers.

Other incarnations of time past inhabit this region: wild horses. The basin is one of the few remaining tracts of Western land where you can still find large herds of them. Like memories of the emigrant trains, they too have been caught in the basin's space warp, protected by the remoteness of the land from the predatory march of civilization.

Perhaps you have never seen a wild horse. This would not be surprising, for in all the United States only about 17,000 are left—not very many considering that half a century ago the herds ran a million strong. Explorers in the 19th Century reported bands of nearly a thousand, and one account tells of watching the herds move without interruption across the horizon for an entire day.

These great herds played a vital part in Western history. Many Western Indian tribes counted their wealth by the number of horses they owned. Heaven to a Blackfoot consisted of horses, women and buffalo —in that order. Indeed, without horses the Plains Indians could never have been so successful at tracking down the buffalo, on which they

Huge holes and crevasses riddle the rain-eroded sandstone of the Honeycombs, a unique 20,000-acre area within the Great Divide Basin.

subsisted, or raiding wagon trains or holding out against the United States Army for as long as they did. Lewis and Clark packed through Idaho with horses purchased from the Indians, as did Stuart, until the Indians stole his back from him.

But in recent decades wild horses have been systematically hunted down and destroyed, expelled from their range to make way for livestock, run down and shot at from jeeps and small airplanes, their carcasses sold for dog food at six cents a pound. Fortunately Congress has passed a law protecting wild horses, surely one of the more enlightened actions that body has taken.

There is a certain nobility about a wild horse that a saddle animal never quite reaches. As he gallops free across the plains, his mane tossing, his tail pluming behind him in the wind, he seems to belong to some special aristocracy of nature, some primal embodiment of speed and power. No other animal, wild or domestic, has exactly this quality, an intangible *esprit* that, like quicksilver, slips away when you try to catch it. Confined to a stall, a wild horse can seem a mangy and dejected creature. The stallions frequently die when caught, as if pining away for their lost liberty. On the other hand, here is what a taste of freedom can do for a domestic plug, from an account by a Rocky Mountain explorer named David Thompson in 1809: "A dull, mere pack horse was missing and, with a man, I went to look for him and found him among a dozen or so wild horses. When we approached, this dull horse took to himself all the gestures of the wild horses, his nostrils distended, mane erect and tail straight out."

Some people will tell you that wild horses are not really wild. And there is truth in this. The original horses of the American West, from which the wild herds are descended, were domestic stock brought in by the Spanish in the 17th Century. These were desert breeds, Arab and Barb, that had come to Spain from North Africa during the Moorish conquest. The Spaniards called them *mestengos* (later Americanized to mustang). Tough and sinewy, with slender legs and small compact bodies, they possessed an almost uncanny ability to travel long distances on little feed and water. The Spanish used them not only as transportation but also as instruments of awe and conquest against the Indians. Suitably impressed, the stronger Indian tribes realized that the best defense was to get some horses of their own. This they did, by bartering skins for them and by riding the Spanish herds. And thus they developed the horse-borne buffalo culture that prevailed throughout the Great Plains and much of the Rocky Mountains.

Many of the horses, both Indian and Spanish, broke loose and ran wild. At the same time settlers coming from the East brought their own breeds of horseflesh, and some of these also ran away to mingle with the wild bands. The mustang strain became further diluted by those of the garden-variety plow horse, of the United States Cavalry remount herds and any other equine tramps that happened along. So today a purebred mustang is a rare creature indeed.

But to most Westerners wild horse has always meant mustang. And the mustangs flourished in the arid plains landscape as few other animals could. Today more than 2,000 mustangs—purebred or otherwise—live in the divide basin and the surrounding hills. It is the largest herd in the country. But the basin covers a lot of territory, and as we drove toward the Oregon Buttes, scanning the hillsides with field glasses, we did not spot a single horse. Even by the time we had reached the top of the first butte and stood looking out over the diminishing perspectives of the basin itself, no mustangs had appeared.

Other animals lived on the butte. We had abandoned the truck to climb the butte on foot, through the only grove of pine trees within scores of miles. A bluebird flew up from his nest in a hollow trunk, his feathers a powdery azure in the clear light. It seems a bluebird's color is pure illusion; he is not really blue at all. There is a legend of the Paiute Indians that explains this paradox with great charm and a certain degree of scientific accuracy.

Ages ago, before there were any birds, the Great Spirit would suffer a mild gloom each autumn because the trees were about to lose their beautiful foliage. One year the Great Spirit decided to do something about it: he turned the leaves into birds. The russet oak leaves became robins, the yellow aspen leaves goldfinches, and so on. But by some divine oversight the bluebird received no color. Bitterly disappointed, the bluebird flew up to heaven to protest, and en route he blundered through the curtain of the sky. Bits of the sky rubbed off on him, leaving him as pretty a blue as any bird could wish.

And that in a sense is what really happens. A bluebird's feathers contain no blue pigment. Their color comes entirely from their structure, which includes an arrangement of tiny cells that scatter and reflect blue light. The sky itself gets its color in virtually the same way; tiny dust particles and water droplets scatter the blue light rays and turn the atmosphere a rich and celestial blue.

After an easy half-hour climb Tom and I stood on the rimrock at the

Rutted tracks worn down by countless thousands of covered wagons still mark the gentle hills at South Pass, Wyoming. The relative ease

of crossing this portion of the Great Divide made it a favored 19th Century route for Oregon-bound settlers and California gold seekers.

crest of the butte. There we had a front-row balcony seat over the vast and disappearing spaces of the desert below. A brownish haze filled the basin—the effect of all those empty distances—obliterating the horizon and softening the profiles of the land forms. Yet we could make out the bluffs along the northern rim, creased and gutted by erosion into fanciful shapes, the weird desert architecture of mesa and arroyo. But still no horses.

We camped for the night at the foot of the butte, and next morning we headed into the basin. The wind continued strong, sending tatters of gray cloud streaming overhead, with islands of shadow trailing behind them across the hillsides. We climbed a small rise and studied the land to the south. Less than a mile away on the slopes of the next hill we saw a dozen horses grazing: a roan stallion and his harem of mares.

There is a system of manners and precedents in the horse world as rigorous as the protocol at a Byzantine court. Its basis is unabashed male chauvinism. Mustang stallions, like certain Oriental potentates, measure their status by the size of their harems. After adolescence they spend most of their time rounding up concubines, bullying them into line and guarding them jealously against all challengers—most specifically other stallions. Single stallions continually try to raid the harems. Fierce battles develop, with lots of biting and kicking, until the challenger either deposes the harem owner or retires in defeat.

When left to himself a mustang stallion will dominate his mares for the sheer joy of showing who is boss, herding them the way a collie herds sheep, ordering them to run and wheel, nipping the flanks of any who do not obey quickly enough. At the first hint of danger—a predator, another stallion, a scent of man—he will round them up into a bunch and stand facing the intruder in an attempt to stare him down. Then he may spin around and gallop away, driving his harem in front of him across the plains. The same mare always takes the lead—the number 1 wife, so to speak.

As we watched the grazing mustangs, the stallion must have noticed us against the skyline. He clearly did not like having us so close. He must have signaled his mares, for they began moving nervously and bunching up. Then they were off, trotting single file up the ridge.

We drove down to the floor of the basin and out across the sand and mud flats. For an hour or so we saw nothing, no animals of any kind. The sky had cleared, and the sun baked down so fiercely that the air shimmered. Then far in the distance, punctuating the haze, a cluster of black dots appeared. They could have been boulders, except that boulders do not move. We counted a dozen, then 15, then 25. As we drove closer, other dots condensed from the haze, six here, eight there, so that by the time the dots had begun to take on shape and color our tally had reached 100. And still the numbers climbed . . . 130 . . . 185 . . . 210 . . . 234. Two hundred and thirty-four wild horses!

Not even Tom had ever come across so many at once. And in a dip in the land, writhing and shifting in the heat like a mirage, we found the reason: water. The horses had come to drink. A truce apparently prevailed among the stallions, and the herds watered in peace. The truce seemed to include our truck, and we were able to drive past the outlying bands before the horses took much notice. Not until we were well among them did the closest stallions start rounding up their harems. The mares trotted edgily into the background and the stallions turned to face us, pawing the dust and bristling. Then all of a sudden, by some common trigger of instinct, every horse in sight began to gallop.

A strange feeling comes over you as you watch a wild horse run. A tremor moves up your back, your pulse quickens and you want to run after him. The feeling is electric and irresistible, even with a single horse. And here was this immense herd of them, stampeding all around us. Tom jammed down the accelerator and we lurched forward, bouncing over ruts and hummocks.

Why these herds should be so exciting, so touched with magic, I do not know. Perhaps the best explanation comes from a trail hand named Matt Field, in 1839: "He bounds away . . . swift as the arrow of the Indian's bow, or even the lightning darting from the cloud. We might have shot him from where we stood, but had we been starving, we would scarcely have done it. He was free, and we loved him for the very possession of that liberty we longed to take."

And freedom is what a mustang is all about. As he moves across the landscape—neck arching, muscles working to the lilt and rhythm of the gallop—he becomes a kind of talisman for everything wild and beyond our reach, an essential distillation of wind and sun and grass and indeed of the wilderness itself.

In any case, we could not keep up. We watched the horses gallop off and disappear into the haze, until only their dust remained.

The Hunters and the Hunted

Among the West's most prized natural resources during its days of discovery were its magnificent animals. Four of them especially—the grizzly, the bighorn sheep, the mountain lion and the elk—were avidly hunted by the rich and famous as well as by the humblest frontiersman; and all four species were driven to the edge of extinction. But now, protected by law and man's increasing awareness of their worth, all of them are making a comeback and still grace the wild parts of the Great Divide country.

In the early 19th Century most of these animals were known to Easterners and Europeans only through the intriguing descriptions of explorers. But when the Western frontier opened, big-game hunters, their appetites whetted by tales of spectacular gunning, began to journey west in search of trophies. One of them, the Irish baronet Sir George Gore, arrived in Colorado in 1855 with 40 attendants, 14 dogs, 112 horses, 6 wagons, 21 carts and 12 yoke of oxen. This extravagant three-year expedition resulted in the slaughter of some 40 grizzlies, 3,000 buffalo, and deer and antelope beyond count.

In the case of grizzlies, the settlers could only applaud the carnage. *Ursus horribilis* had become the nemesis of their cattle operations. Some bears left such identifiable tracks during their raids on the herds that ranchers came to recognize them and give them nicknames such as Old Clubfoot *(right)* or Three Toes or some other descriptive. For grizzlies in general, an all-inclusive epithet seems to have been, for some reason, Old Ephraim.

Grizzlies are indeed awesome adversaries. A big one stands 10 feet tall on its hind legs and can fell an animal as large as itself with one incredibly swift blow of its paw; a grizzly was once seen to kill four bull bison in a single attack. Understandably, the grizzly has no foe in the animal kingdom except a man with a rifle and steady nerves.

What threatened the grizzly was man in the mass, with his civilization encroaching on its former preserves. Only in still-unsettled areas like Alaska do they number in the thousands. In Montana, Wyoming and Colorado there may be only about 800 left. But now that they are protected by law, the chances are that there will still be a few grizzlies in the Great Divide country as long as there is some wilderness left for them to prowl.

The most famous bearskin rug of its time, displayed at the St. Louis World's Fair of 1904, was originally known as Old Clubfoot, a grizzly notorious for its depredations on Colorado cattle ranches, and nicknamed for the three missing toes on its right front paw. In its palmier days Old Clubfoot sported the same thick coat and characteristic hump behind the neck seen on the female grizzly in the inset.

Bighorn: Wary Lord of the High Places

For many big-game hunters the bighorn sheep *(inset)* is the most challenging quarry: the sheep largely live above timberline and are so adept at jumping from one ledge to the next that only the most nimble-footed Nimrod can give chase.

Before the great open spaces of the West began to close in, the bighorn not only were more plentiful, but also lived in lower, less precipitous terrain. Indians in the Yellowstone area used to capture the sheep in large numbers, and relied on them as a major source of food. Few people get a chance to taste bighorn meat today; as the animals became scarcer, protective state laws were passed that now permit only a few of the sheep to be taken each year.

So, though diminished in number and restricted in range, the bighorn are thriving in their rarefied habitat. Lambs are born in early spring and live with their mothers in flocks in the high country. The rams lead a wandering bachelor life most of the year but, come December, return to the ewes to mate, and to fight battles in which two rams competing for a female duel in thunderous horn-to-horn combat. But for all their formidable prowess, bighorn are the most wary of animals where people are concerned. The merest glimpse of a human will instantly send them bounding over the rocks and out of range; and this wariness is perhaps as important a factor as any other in their continued survival.

Girded in his cartridge belt and holding a .30-caliber Winchester

rifle, a Montanan of the 1890s poses proudly with the three bighorn rams he has bagged—a rare feat among hunters of these elusive sheep.

Montana hunters of the late 1800s, a period when the only good mountain lion was a dead one, display their handsome victim before taking it

in to collect the bounty then offered for such varmints.

The Mountain Lion: A "Varmint" Redeemed

Sometimes called the cougar or the panther, the mountain lion has always had an ambiguous reputation. The Inca of Peru called it puma, admired it for providing a sporting chase and hated it as a predator of livestock. The Indians of Baja California viewed it with superstitious gratitude because they often depended on the leftovers of the animal's feasts for their own sustenance. To sheep and cattle ranchers, however, the mountain lion of the American West *(inset)* has always been the enemy, a sneak thief in the corral and on the range; until only a few years ago many Western states placed a bounty on its head.

There is no doubt that mountain lions do indeed seize an occasional lamb or calf and they are accomplished deerslayers as well. Weighing as much as 200 pounds, they are the second-largest cats in the Western Hemisphere—only the jaguar of South and Central America is bigger—and they have the big cats' natural endowments of speed and strength to attack and fell the young deer and elk that are their favored prey.

In most places, however, the cougars' numbers have been so reduced by trapping, poisoning and shooting that they can no longer do much damage to sheep or cattle. Gradually their former image as hated varmints is fading. The last state to offer a bounty on cougars rescinded it in 1971, and there are signs that America's big cats are making a comeback.

The Elk: Target for Food as Well as Sport

Once the most widespread species of deer in North America, elk were found as far east as New Jersey and as far south as Alabama. But it was in the high country beyond the Great Plains that the elk's size—it can weigh half a ton—and regal crown of antlers *(inset)* evoked particular admiration: more than one 19th Century writer acclaimed the animal as the "monarch of the West."

Such literary cachet did not protect the elk from being nearly exterminated as an advancing tide of settlers hunted them for food and trophies. Many frontier communities found elk to be the cheapest, most convenient source of meat, and to satisfy this demand professional hunters killed off many herds.

In this century, the elk has been a tempting target for trophy hunters seeking the male's imposing rack of antlers as well as for those pursuing a more curious prize: the elk's two upper canine teeth, or tusks. For a time members of the well-known fraternal order named after the elk affected the vogue of wearing a tusk on their watch chains, and considerable numbers of elk were slaughtered to cater to that fashion.

Protective measures have slowed down, and in some cases reversed, the dwindling of the herds. In Yellowstone Park the elk are coming back so fast that park officials have shipped thousands of them, by truck and train, to repopulate wilderness areas in other states.

Citizens of Clear Creek County, Colorado, relax amid their

paraphernalia during a hunting expedition for elk and other deer. Such forays were frequent in the 1890s to secure meat for mining families.

5/ Back and Beyond in Yellowstone

Deer walk upon our mountains, and the quail
Whistle about us their spontaneous cries;
Sweet berries ripen in the wilderness....

WALLACE STEVENS/ *SUNDAY MORNING*

A crowd was already gathering. Most places had been taken on the long horseshoe bench at the edge of a broad apron of cinders, and late arrivals were bunching up on the cement walk behind it. There were people of all ages and sizes and sexes, men in shorts and flowered sport shirts, men in jeans and hiking boots, young women with babies, blue-haired matrons with harlequin glasses, bicyclists with rucksacks, photographers with instant cameras, other photographers with enough expensive camera gear to open small shops. Everyone was waiting as though for a curtain to rise or "The Star-Spangled Banner" to be played, and everyone was asking a question.

"Hey, when's it gonna go off?"

"Couldn't say, we're just standing here like all the other folks."

The sun boiled down on the cinders and a strange odor, as though a shotgun had been fired, tinged the air. Several people licked ice-cream cones bought from the food concession back under the pine trees next to the ranger station.

"How sooooon, Mommy?"—a nagging, exasperated sigh.

"Now just you stand quiet and watch, son."

The object of everyone's attention was a grayish mound shaped like a small volcano, about 200 yards in front of the horseshoe bench. The mound was squat and rather ugly, its sides rough with mineral deposits, and from time to time wisps of vapor would float skyward from the crater at its top. Many of the people had driven across half the country to look at it. For this ash-gray hummock is one of the most celebrated natural sights in North America, more thoroughly gaped at, photographed, publicized and written home about than any other item on a tourist's calendar. It is a geyser called Old Faithful.

Here in Yellowstone National Park, where the Continental Divide cuts through the northwest corner of Wyoming, is the greatest concentration of geysers anywhere in the world. At least 200 of these natural water jets fume and spew in Yellowstone, along with several thousand hot springs, boiling mudholes, vapor ducts and other thermal curiosities. And Old Faithful is the star performer of the lot.

Old Faithful is Yellowstone's most famous geyser for a particular reason: unlike many other natural steam jets, it never misses an eruption. The intervals between eruptions vary somewhat, but roughly once every hour Old Faithful's pressures build, its plumbing begins to boil and more than 10,000 gallons of superhot water goes shooting into the air. And this performance has never been known to fail since the geyser was discovered more than a century ago.

Everyone was checking his watch, studying National Park Service brochures that explain the mysteries of geysers *(pages 148-157)*, and running to the Visitor Center where the estimated time of the next eruption was posted on a bulletin board. At exactly 1:46 on this particular afternoon, the crater gave a low burp and a dollop of water sloshed over its edge. The crowd mumbled uneasily and loosed several catcalls. "You've kept me waiting here for that?" one man said to the lady next to him, and turned to leave. But Old Faithful had simply cleared its throat. Suddenly it let out a whoosh, and sent up a dense plume of steam and hot water about 100 feet high. The downwind spectators stampeded for shelter as the hot spray drifted over them. The rest of the audience clapped and cheered while the fountain rose still higher, poised like a ballet dancer on tiptoe and then subsided. The performance lasted about five minutes. When it was over, everyone started walking back to the parking lot, ready to drive on to the next famous natural wonder down the smoothly paved road.

The first men of historical record to witness Old Faithful in action did not treat the event so lightly. They were in a party of Montana citizens led by the state's surveyor general, Henry Dana Washburn, and they arrived at this geyser basin in September of 1870 after trekking through the Yellowstone wilderness for almost a month. The men had traveled up the Yellowstone River from Montana and had seen some rather impressive sights on the way: a huge canyon more than 1,000 feet deep in places, a pair of immense waterfalls and any number of boiling springs and sulfurous mud pots. After picking their way through the pine forests that surround Yellowstone Lake, they emerged into the geyser basin just as Old Faithful was erupting:

"We had within a distance of fifty miles seen what we believed to be the greatest wonders on the continent," wrote one awed member, Nathaniel Langford, in his journal. "Judge, then, of our astonishment on entering this basin, to see at no great distance before us an immense body of sparkling water, projected suddenly and with terrific force into the air to the height of over one hundred feet. We had found a real geyser." So entranced were the Montanans that they stayed for nine eruptions.

Shown close up, the lichen-speckled surface of Obsidian Cliff in Yellowstone Park reveals the tortuous patterns left by molten rock that congealed quickly to produce obsidian, or black volcanic glass, and lighter-colored veins of quartz and feldspar.

Before the Washburn safari, very few people believed that such wonders as Old Faithful really existed. True, rumors of weird and diabolical phenomena near the headwaters of the Yellowstone River had been circulating for more than 50 years, brought back by various rude and buckskinned mountain types. John Colter, a member of the Lewis and Clark expedition who had stayed in the Rockies to trap beaver, returned to civilization in 1810 with an outlandish tale of gigantic boiling springs and other sulfurous waterworks somewhere in the mountains. He also described an escape from Indians that, if true, was downright miraculous. It seems Colter had been caught by a war party of Blackfoot, who decided to have a little fun before putting him to death. They removed his clothes and moccasins and told him to start running. Colter lit out, stark naked and unshod, with no fewer than 500 Blackfoot warriors in bloodthirsty pursuit. Colter claims to have outpaced all 500—except for one, whom he turned on and killed barehanded. Everybody believed Colter's Indian story; but boiling springs and brimstone? Such things were too fantastic to be taken seriously.

Even harder to swallow were the stories of another old mountain man, Jim Bridger, who wandered through Yellowstone in the 1820s or thereabouts. Bridger was a respected guide and trapper—he was the first white man to see, among other things, Great Salt Lake. But Bridger also liked a good yarn.

There was a huge lake, Bridger said, at the top of the Yellowstone Plateau, where for a depth of several feet the water was scalding hot; you could catch a trout and by the time you had pulled it to the surface the fish would be cooked. Bridger also told of a creek so full of alum that it shrank everything, including time and distance. He claimed to have found a canyon so large that before retiring he would yell "Wake up, Jim," and be roused the next morning by the echo. He described a mountain made out of glass and a forest of "peetrified trees a-growing, with peetrified birds on 'em a-singing peetrified songs."

Fantastic, yes, but all with a grain of truth. Parts of Yellowstone are indeed littered with petrified stumps and limbs, the sun glints off cliffs of a black volcanic glass called obsidian, and the Grand Canyon of the Yellowstone is indeed one of the country's most prodigious gorges. In one part of Yellowstone Lake, which sits in the center of the park only three miles from the Continental Divide, hot springs do occasionally cause the surface water to become quite torrid. Not hot enough to poach a trout perhaps, but that can be accomplished in another spring nearby.

The stories persisted, but for years nobody bothered to investigate them. And for good reason: the Yellowstone region was all but impossible to reach. Mountain ranges hem it in like the walls of a stockade, and snow clogs the passes for a full nine months each year. Even the Indians, who called the region the summit of the world, went there only when they had to, in order to gather pigment from the mineral deposits around the hot springs or to chip arrowheads from the obsidian cliffs. And so Yellowstone remained terra incognita.

But mysteries cry out for solutions and eventually a few adventurous souls went in to see what was really there. In 1869 three Montana mineworkers bushwhacked as far as Yellowstone Lake and stumbled upon a fair number of hot springs and geysers. Two of the men, Daniel E. Folsom and Charles W. Cook, had acquired a bit of schooling, and they took note of these features with a clear and educated eye. Even so, their account, to a reader today, seems frankly surrealistic. Here is one sample from Folsom's diary:

"We followed up the Madison five miles, and there found the most gigantic hot springs we had ever seen. They were situated along the river bank, and discharged so much hot water that the river was blood warm a quarter of a mile below. One of the springs was two hundred and fifty feet in diameter. . . . There were no fish in the river, no birds in the trees, no animals—not even a track—anywhere to be seen, although in one spring we saw the entire skeleton of a buffalo that has probably fallen in accidentally and been boiled down to soup."

Folsom and Cook combined their observations and submitted them for publication to various magazines. "Thank you," replied an editor of *Lippincott's Magazine* in Philadelphia, "but we do not print fiction."

But in Montana, close to the region itself, people bought the men's story. And so the next year, in late summer, 1871, Henry Washburn, with eight other prominent Montana citizens and an Army escort, headed into the Yellowstone country on horseback to make a thorough exploration. The party stayed 42 days, trailing through country so rough that their mascot, a dog named Booby, had to be fitted with moccasins. They circled Yellowstone Lake, scrambled up the park's most prominent mountain (which they named Mt. Washburn), discovered—and christened—Old Faithful, measured the temperatures of the hot springs, brought back samples of the water as well as various mineral specimens to be analyzed in a laboratory. They also brought back some deep and powerful feelings about what they had seen. "Astonishment and wonder," wrote Nathaniel Langford, "become so firmly impressed upon the mind . . . that belief stands appalled."

Astonishment and wonder. Strange words from a member of the Washburn expedition, which consisted of hard-headed business people and state officials. Some of them saw immediately the commercial potential of the Yellowstone region, and talked of tourist concessions and resort hotels. They came out with a different idea entirely. Around the campfire at the end of the trip a remarkable upsurge of altruism took hold of these men and gave birth to a new concept in the way this country treated land. No private speculator should be allowed in Yellowstone. Its marvels must belong to everyone, preserved in perpetuity for the entire nation.

With the enthusiastic backing of the whole group, Langford set off for Washington, D.C., where he badgered the government into conducting a federal survey, led by the influential Ferdinand V. Hayden. Photographs were taken, maps drawn and reports compiled, and the combined result was placed on the desk of every senator and congressman on Capitol Hill. Congress reacted as Langford had known it would—with wonder and astonishment. Legislation was passed, and on March 1, 1872, President Ulysses S. Grant signed the bill creating Yellowstone National Park as a "pleasuring-ground for the benefit and enjoyment of the people."

Yellowstone National Park is now in its second century, and to someone caught up in the sightseeing throngs on a midsummer day it seems that all the people in America have indeed come here to gawk. Some 2.5 million tourists pour through the park each year—an average of 30,000 a day during the summer season. Most of these visitors spend their time driving along the paved highway that connects the park's

The diminutive algae-covered cone of Tom Thumb Geyser rises about one foot from the base of the Grand Canyon of the Yellowstone.

carefully signposted scenic attractions—Old Faithful, Tower Falls, the Grand Canyon of the Yellowstone. Few ever leave their cars. Those who do generally restrict their modest explorations to the hot springs and geyser basins, where they amble obediently along the wooden walkways constructed by the park service to keep people from breaking through the crust and scalding themselves. As the visitors cluster around the springs in little groups, bending in rapt attention over the vents, with the steam rising into their faces, they resemble picnickers roasting hot dogs at a barbecue pit.

Then they climb back into their cars and again follow the highway signs in search of wild animals. The park's billboards dutifully advertise the favorite habitat of each species. Sure enough, the animals are there, for Yellowstone contains one of the richest concentrations of wildlife anywhere in the United States. An estimated 17,000 elk graze in the park's plateaus and meadows, along with 700 buffalo, one of the world's largest surviving herds. There are mule deer, black bears and grizzlies, bighorn sheep, coyotes, mountain lions, eagles, ravens, plus any number of waterfowl, including swans, egrets and pelicans.

Almost invariably, when someone spots an animal, he slams on his brakes and leaps out of his car to have a look. Traffic backs up behind him, and the other motorists start shouting and leaning on their horns just as though they were driving on the Los Angeles Freeway. The road signs also warn that big animals are dangerous—keep away from them. But people constantly disregard the signs and run the risk of getting mangled. I happened upon one traffic jam caused by a large buffalo grazing at a roadside. He was a great hulking beast with menacing horns, and all the people had left their cars to take his picture. One young man had edged to within 10 feet of him and stood there focusing his camera. The buffalo showed more patience than I would have; he simply turned his posterior to the camera and kept on grazing.

The highway circuit undoubtedly has its merits. It is quick and convenient and provides perhaps a teaspoonful of wilderness concentrate for millions of people who otherwise would never get so much as a taste. But something essential is missing. With all its instructional signs, scenic overlooks and brief nature trails through the geyser basins, it lacks the quickening undercurrent of excitement that runs so strongly through the journals of the early explorers. Where is the wonder, where the astonishment? Where is the sharp and unmistakable pang, the catch of breath and flutter of pulse that come when you discover, by yourself, something completely new and unexpected?

Most regal of North American birds, the golden eagle—named for the color of its head feathers—is three feet tall and has a seven-and-a-half-foot wingspan. It soars with superb grace and can dive at 150 miles an hour when swooping to seize its prey, mainly rabbits, snakes and various rodents.

Just beyond the highway, a 10-minute walk from the Old Faithful parking lot, a hundred yards off the macadam through the first row of trees, wonder and astonishment are waiting. There are 3,742 square miles of land as clean and untouched as it was a century ago; elk and buffalo graze in privacy and the grizzly bear lumbers unnoticed, a great furry ghost among the pines and huckleberries.

I went looking for them. Like many of the early explorers, I had decided to go on horseback. One morning in late August I stood at a trail head on the east shore of Yellowstone Lake, tightening the cinch against the swelling belly of a chestnut mare named Derby. Holding Derby's bridle for me was a friend, Sim Smith, who had flown in from New York with his 15-year-old son, Aimery. Our guide, an old Montana hand named Hank Rate, loaded gear onto two packhorses—sleeping bags, two tents, an ax, extra horseshoes, cooking equipment and provisions for five days. Our plan was to travel south along the lake to its upper end, then head west into the least-traveled section of the park, crossing the Continental Divide at a place called Two Ocean Plateau. In imitation of Henry Washburn and Nathaniel Langford, who traversed this same country, I decided to keep a journal.

Tuesday, August 29: Overcast day, raw and autumnal, with a threat of rain. As we truck the horses to a trailhead we spot several dozen Canada geese, pausing to feed on their way south along the autumn flyway. "Looks like it's going on fall," Hank says. He tells us that it often starts snowing in Yellowstone at this time of year.

The trail begins halfway up the east shore of Yellowstone Lake in a lodgepole-pine forest. We saddle our horses, tether them to saplings and give a hand to Hank. He loads the packhorses, a painstaking job since the loads must be carefully balanced to keep them from shifting. Gear and provisions are stuffed into paniers, pairs of boxlike compartments joined together by canvas straps. The straps are draped over the horse's back so that a compartment hangs on either flank. Hank gauges the weight in each compartment with all the delicacy of a man assaying gold dust. He removes a box of pancake mix from the right-hand side and transfers it to the left, then hoists the panier onto the horse.

We set out at about 11 a.m. through the lodgepoles, Hank, elegantly mounted on a beautiful high-stepping mule called Festus, leading the pack animals. The rest of us fall in behind, single file. The forest seems quite lush, with a lacy ground cover of dwarf huckleberry, which has tiny leaves and diminutive purple berries. In the glades we identify purple aster, wild parsley, the blue flowers of lupine. Mushrooms sprout everywhere—purple, white, crimson, scarlet, a sign of recent rain. From time to time the forest opens into meadows, allowing a clear view across Yellowstone Lake to our right. Above us to the left we catch sporadic glimpses of rock domes and turrets—volcanic outcrops of the Absaroka Mountains, which seal off the park's eastern boundary.

We have left the geysers behind us to the north, but every now and again we catch a whiff of sulfur from some unseen hot spring back in the woods. In one place the lodgepoles thin out and all other vegetation ceases, so that only brown pine needles cover the ground. As we ride across this stretch the earth reverberates as though the hoofs of our horses were beating against the head of a giant drum. In a sense, they are. According to Hank, the ground underneath is hollow. Hot springs once flowed here, he thinks; their craters are now sealed over with a crust of mineral deposits and the soil is still too poor for most plants.

As we ride on, an unspoken competition develops: which of us will see the first sign of game? Naturally it is Hank. He points to a pile of animal droppings on the trail, brown pellets the size of marbles. "That's mule deer, bigger than a rabbit and smaller than an elk," he says, and goes on to explain that game managers have an intriguing way of using these piles—known in the profession as scat—to estimate the mule-deer population on a range. It seems that each deer, by the peculiar mathematics of nature, defecates 13 times a day. By counting the piles of scat in a certain area and applying some arithmetic, a ranger can determine the number of deer per acre. But for an accurate count he must be careful not to include last year's scat.

Hank is something of an original. He has followed perhaps a dozen professions—forest ranger, ranch manager, cowboy, surveyor, blacksmith, elk hunter, horse wrangler, freelance journalist—and now he owns a small farm in Montana. He is pungently articulate in an Old West kind of way, which means that he does not talk, but palavers, and says things like "We're fixin' to get some hellacious weather tonight, when those ole thunderheads come a'pounding and resounding over the ridge there." Such turns of phrase would sound corny and contrived when delivered by a lesser man, but Hank carries them off with an easy style that is totally convincing.

Farther down the trail we see another kind of animal droppings. Bear. Nowadays few park visitors ever see a bear, despite the cliché photographs of Yellowstone bears panhandling food along the highway.

These animals by nature have always been solitary creatures who hate the very smell of humans. In past years, however, so many people have fed the bears that the animals have overcome their aversion to get the handouts. Happily, bear feeding is now strictly against park rules and the bears are returning to the woods. This is a good thing, for even black bears, cuddly and lovable as they seem in pictures, do turn vicious. In the past 25 years 1,149 Yellowstone visitors have been injured and one killed by black bears. And then there are the grizzlies.

You do not fool around with a grizzly bear. A full-grown adult weighs almost half a ton, can stand six or seven feet tall on his hind legs, and for all his bulk moves with surprising agility. He can run a man down at a speed of almost 40 miles an hour. About 250 grizzlies now live in Yellowstone, according to the latest count, and there is always a chance here in the back country that we may find ourselves face to face with one. People hiking in this area often attach bells to their backpacks on the theory that the bear will move away from the noise. But grizzlies are notoriously unpredictable, and one Yellowstone naturalist claims that bear bells may simply arouse the beast's curiosity, causing him to come and investigate. No one really knows.

One thing is certain, however: grizzlies are almost always hungry. They will go to any length of assault and burglary to get a meal. We therefore take certain precautions with our provisions. Bacon, cheese and other strong-smelling foods are packed in airtight containers. All refuse is either burned or stuffed into airtight bags. Edibles are hung from a tree limb above any bear's reach and well away from the tent. No one who follows these rules has trouble in Yellowstone, but woe to the man who disregards them.

Earlier this year two young men camped in the woods near Old Faithful, at a spot that had been put off limits because of the bears. Their campsite was a rural blight of dirty clothing, open provisions and half-eaten sandwiches. One night the pair returned from a late party at the Old Faithful lodge to find a grizzly riffling through the mess. One lad shone his flashlight in the bear's face, and the animal attacked. The bear dragged him into the bushes, clawed him to death and proceeded to eat part of his torso. His companion fled, hysterical with shock, to get help at the Old Faithful ranger station. The rangers hunted down the bear and shot it. The animal turned out to be an old sow with most of her teeth missing, no longer able to feed herself in the wild. She had discovered an easy new food supply and was simply defending it.

Hank studies the bear scat on the trail and says it comes from either

The Yellowstone River plunges 308 feet down Lower Falls into the Grand Canyon. Above the falls, hard volcanic rock has resisted the water's scouring; below, an area of once-similar rock, "rotted" by thermal activity from underlying magma, has yielded to the river's erosion, creating the canyon and exposing the cliffs whose color gave both river and park their name.

a large black or a small grizzly. "Don't worry," he says, "the scat is several weeks old." I am disappointed, for I would very much like to see a grizzly—from a distance of course. Everyone tells me I'm nuts.

We ride on without further incident, the pines alternating with fine grassy meadows where we gallop our horses. By late afternoon we reach some bluffs overlooking the upper end of Yellowstone Lake—our first campsite. Ahead we can see a stretch of marsh called the Thorofare, where we will ride tomorrow. Its grass and sedges are a bright burnished gold wherever the sun breaks through the overcast sky. A network of game trails crisscrosses it, interrupting the texture of its surface like stitching across velvet. Elk or moose, according to Hank. "I can about guarantee you'll see moose there tomorrow," he says. Across the Thorofare and a dozen miles to the west we can make out Two Ocean Plateau—named for a swampy area precisely on the Continental Divide where two streams run together and then split so that one branch flows to the Pacific, the other to the Atlantic. Above us a snowbank from last winter hangs from the rimrock like a swag of white drapery.

We set up tents near the edge of the bluff, build a fire and cook dinner. A brief shower drives us into our large three-man tent. But the rain quickly passes and the sky clears. A fine warm evening. I now sit at the edge of the bluff near the fire with my back to a tree. The moon climbs out of the forest and pours its calm and generous light across the water of Yellowstone Lake.

Wednesday, August 30: Reveille at 6 a.m., with cries of waterfowl from the lake. From the Thorofare comes a soft flutelike whistle—a moose calling. The sun rises, and the morning turns clear and hot.

Today's trail threads in and out of the evergreens, through a tangle of willows in the Thorofare and along the Yellowstone River. "Prime grizzly territory," Hank says. I think he is kidding, just to scare us, when suddenly he reins in his horse and points to a grotesque black shape 20 yards ahead among the willows. False alarm. No bear, but the promised moose, an ungainly creature with a humped back, dewlap and outsized ears. We stare at the moose, and he at us, for about five minutes, hardly breathing. Then the moose trots off into the trees.

Signs of other animals fill the woods. We see trees where the bark has been torn, chewed, scraped or clawed; an experienced woodsman can tell what kind of animal has made each mark. Hank indicates an ominous set of claw marks—deep parallel gouges as though the tree trunk had been slashed repeatedly with a knife. Bear. Apparently these animals compete to see which can reach the highest and scratch the deepest. One set reached 10 feet above the ground and penetrated more than an inch below the bark. Only a grizzly could have made them.

Some trees show bare patches that from a distance resemble the ax cuts used to blaze a trail. Closer inspection shows them to have been caused by deer and elk rubbing against the trunks to clean the velvet from their year's new growth of antlers. Antlers are beautiful things —great branching crowns of bone—and they have always fascinated me. The antlers on a mature bull elk may weigh 25 pounds and span five feet. Every year a bull drops his antlers in late winter and then must grow a whole new set. The growth process takes most of the summer and is painful, in the way that an infant feels pain when he cuts new teeth. New antlers are soft and sensitive, covered by a layer of velvety skin that is full of nerves and blood vessels. By August the antlers have hardened into bone and the velvet can be scuffed off. The animals pick saplings, which have a bit of give to them. An elk mark is typically about five feet high on a sapling several inches thick. Buck deer, being smaller, pick thinner saplings and rub lower down.

A bull elk's antlers (like those of buck deer) have only one use: to battle other males during the autumn rut. In Yellowstone this begins early in September and continues through October. The males that have spent most of their time running together in amicable bachelor groups now turn mean. Their neck muscles swell, they forget to eat, and they start tearing up the sod with their hoofs and antlers. "Whomping and stomping and getting mad enough to fight snakes," is the way Hank puts it. The bulls send a war cry to each other, a melancholy, high-pitched bugle that sounds like a jazz cornet in the upper registers. Then the battles begin. All to see which bull mates with which cows.

Two Ocean Plateau is the summer range of one of Yellowstone's two elk herds. We hope when we reach it to find that the rut has begun. And so we cut west across the Thorofare and ford the Yellowstone River. The plateau rises steeply from the river's west bank, a climb of about 1,500 feet. We work our way up through a forest of Engelmann spruce, following a narrow trail along Lynx Creek. The weather has closed in hard, and by the time we have reached the top the afternoon has turned grim and dark. A cold rain starts and quickly turns to hail. The pellets sting like buckshot, pummeling our hands and drumming on the hoods of our ponchos. We pick a campsite in a grove of the tallest pine trees I have ever seen, and set up the tents and cook dinner, miserably wet, cold and exhausted.

Thursday, August 31: How glorious it is to wake up to the sunshine. I break a crust of ice in the water bucket and start brewing coffee. We are camped close to the Continental Divide at about 9,200 feet, or some 600 feet below timberline. The plateau landscape rolls gently, the elevations so consistently the same that without a map the divide itself cannot be distinguished among the other rises. Clumps of spruce alternate with subalpine meadow, curtailing any vistas of the mountains that surround Yellowstone.

We ride south through the meadows, looking for elk. Their sign is everywhere. Elk trails are worn into the grass and at a water hole near the divide hoofprints have churned the bank into chocolate fudge. Elk bed down in timber during the day, moving at dusk into the meadows to graze. We study the edges of the spruce groves. Within the first hour we spot a cow elk with her calf. Her coat is a soft tan, the color of a light bay horse, with darker brown stockings and neck. As she turns away from us into the spruce, her white rump patch flashes in the sun. Farther on two bulls stand clear of the timber on a ridge about 200 feet to our right. They are stately animals with tawny backs that grade to walnut-brown necks and heads. They carry their antlers, heavy racks branched like candelabras, with all the hauteur and ceremony of two monarchs wearing their crowns.

We ride east to the edge of the plateau where we can at last look out across the full sweep of the park. The marshes in the Thorofare lie green and gold below us, and the waters of Yellowstone Lake spread out like molten pewter in the hollows between the mountains. The saw-toothed crests of the Absarokas rip the horizon far to the east. We dismount, tether the horses and walk to the lip of the plateau. Below us on the headwall a snowbank lingers, its meltwater feeding a small hanging garden of subalpine flowers—Indian paintbrushes, daisies, elephants'-heads, bleeding down the cliff face. We fill our hats with snow, melt it in a cookpot for coffee and eat our luncheon sandwiches.

"Do you think we'll see an elk fight?" Sim asks.

"Can't guarantee it," Hank says. "The bulls are still running together fairly peaceable. Come up here in a week or two though, and they'll turn nasty. I've been guiding elk hunts for years, in October usually, and we hear 'em bugling from every ridge, mean as hornets."

For some reason we feel no real disappointment at missing an elk fight. It seems enough to have glimpsed the animals as they grazed quietly in this proud rolling upland. Then as we ride back to camp we

A 60-foot-high rampart of travertine, New Highland Terrace was built up by calcium deposits from Yellowstone's Mammoth Hot Springs.

make a discovery that displaces all other thoughts. Sim has taken the lead, gentling his horse down a steep, heavily timbered slope. He reaches the meadow at the bottom, breaks into a trot, then wheels and stops dead. He points at something in the meadow.

We join him at the meadow's edge. The grass is particularly luxuriant and the flowers plentiful. The place seems idyllic, except for one thing: the floor of the glade is covered with bones. Ribs, vertebrae, thighs, skulls, jawbones, bleached gypsum-white and strewn about the hollow as though we had stumbled upon the aftermath of a war. They are the bones of five dead horses.

It is a chilling, uncanny thing, this token of mortality among the grass and flowers. Our voices grow churchyard quiet. What are these bones doing here? What kind of death caught the five horses together, so far from corrals and barns and feed troughs?

We continue back to camp, trying to puzzle out our discovery. Horses are plains animals, and no heritage of instinct or biology draws them into the high country. They seldom go there except when taken by pack oufits such as ours. In some of the larger expeditions, however, there are so many pack animals that the horse wrangler cannot keep track of them all and a few invariably wander off during the night. Such strays ordinarily find their way back home. But here in Yellowstone the grass is so plentiful that they may simply decide to stay. This is probably what happened. The five horses strayed, then lingered too long in these upland meadows. The snow came and then it was too late. Unable to paw through the heavy snow cover to the grass beneath, the horses weakened and starved, and the bears and coyotes fed on their carcasses. We trail on, without talking much, to the comforting familiarity of our campsite in the pine grove.

Friday, September 1: Early this morning we broke our camp on Two Ocean Plateau and trailed west across the divide into the watershed of the Snake River. We now pause for lunch at the top of a high ridge. Here the vistas open up to the west, affording a clear view of three immense spires called the Grand, Middle and South Tetons; these are unquestionably the three most spectacular mountains in the Rockies.

The Tetons were named several centuries ago by French beaver trappers who thought they resembled a woman's bosom; "teton" is French slang for breast. And it crosses my mind that only a very lonely Frenchman could conjure up the softness of a woman from such jagged promontories—particularly in a cluster of three. Yesterday from the plateau we caught pale glimpses of the peaks through a cloud bank to the southwest. Today from the ridge the Tetons surge closer, breaking in and out of the cloud cover like a thought in the process of being formed. It is impossible, in the swarming cumulus, to distinguish their outlines, to separate glacier from cloud or to tell where the mountains end and the sky begins. Yet the sense of vertical thrust is overwhelming. For the Tetons, unlike most Rockies' summits, are taller than they are broad; they rise discrete and self-contained and they do not suffer from competition by a multitude of surrounding peaks.

We scramble down the ridge, through a dark spruce forest that soon gives way to lodgepole pines and lush meadow. The grass reaches so tall in places that it tickles the horses' bellies. I must continually pull up my horse's head to keep her from stopping to eat. Particularly enticing seems to be a large thistle that grows in open spaces throughout the park. It is called elk thistle—or Everts thistle, in memory of a 19th Century explorer.

Truman C. Everts was a member of the Washburn expedition who became separated from the main party and got lost. He was older than the rest and wore glasses; apparently he fell from his horse, which before running off stepped on the glasses and broke them. Everts had no food, and now very little eyesight. He wandered about semiblind for 37 days until he was found by a rescue party. One thing he could see, because it was so large, was the elk thistle. This became his diet. He would dig up the starchy root, roast and eat it; and thus he escaped starvation. We have seen other edible plants by the score—leafy green bouquets of wild parsley, tart huckleberry, mariposa lily, whose bulb was considered a great delicacy by the Indians. All around our campsite on Two Ocean Plateau grew a large fleshy mushroom, a kind of edible boletus known in rarefied cooking circles as cepe. Farther along the trail we find a bright orange fungus called chanterelle, another mushroom luxury with a faint but exquisite aroma of apricots.

The Yellowstone wilderness is in fact extraordinarily generous to a man who knows the woods. Many of the country's other wilderness tracts are bleak and inhospitable—fierce desert or stark alpine crags. No man, and very few animals, could support themselves in these areas from the land alone. But in Yellowstone a man could live quite comfortably gathering berries and roots, shooting moose and elk if hunting were allowed, and pulling trout from the streams. Thus the park's founders, in setting aside the hot springs and geysers, inadvertently preserved a swath of highly desirable Rocky Mountain real estate.

Saturday, September 2: A late, lazy morning in our last campsite in Yellowstone. We are camped on the bank of Wolverine Creek, a tributary of the Snake River. The horses graze in a field of wild timothy. I walk beside the stream bed through stands of blue gentian, black-eyed Susan gone to seed, clumps of gooseberries. I pick a handful—purple, translucent, tart—that the birds and bears have left.

Wild-strawberry plants grow underfoot, their triplet saw-toothed leaves turned russet with autumn, their fruit long since come and gone. The stream ripples across the pebbles in its bed, breaks into quiet pools, turns back on itself, eddies and waits and continues on.

After a sandwich lunch we mount up for the 10-mile ride that will take us to the trailhead. We file out along Wolverine Creek to its junction with the Snake. Ahead the river winds between low hills with gentle meadows and pinewoods. At the edge of a glade we flush two cow elk grazing in the tall grass. With supreme nonchalance they wander back into the cover of the trees and turn to stare at us as we ride past.

Except for those faint whiffs of sulfur on our first afternoon five days ago, we have seen no new evidence of Yellowstone's hot springs. But now as we emerge from the pines and approach the riverbank, the reek of brimstone is unmistakable. It comes from a half-dozen hot springs and fumaroles—steam vents—along the water's edge. They are all tiny perfect white cones with bright blue steaming water sloshing over their tops. The craters are so minute I could cover any one of them with my hat. Yet there is an odd diabolical quality about them that evokes the descriptions of fire and brimstone in the early Yellowstone journals. It is as though the protective crust of the earth had worn thin and a force from the nether world had pushed its way through. Coming upon them by surprise as we have, seeing them perched so incongruously on the riverbank, we find the effect shocking. Nothing has prepared us for it. No guidebook. No brochure. Only five days in the back country among the trees and animals.

And this is the point of our brief odyssey: the slightest discovery has become an event. For during these five days in the wilderness our perspectives have changed. Our senses have been sharpened by those hours in the saddle, our eyes washed clean in the afternoon rains. The wilderness has given us back our sense of wonder; it has restored our capacity for astonishment. We have spotted perhaps a dozen elk, a lone moose, some claw marks made by a grizzly bear. Not much, really. Driving through Yellowstone by car earlier this summer, I counted more game in a single morning, including a herd of mountain buffalo.

But a man in an automobile is insulated from what he sees, cut off by the glass of his window, the noise of his engine and the speed at which he travels. Even if he leaves his car and walks out onto the grass he is still psychologically removed from natural things. His breakfast at a restaurant, his martini the night before, his bed at a motel, all the comfortable appurtenances of modern life cut him off from the land and its animals. They become curiosities, exhibits in an open-air museum maintained by the National Park Service. But a few days in the wilderness strips away the insulation with surprising alacrity. A man's instincts grow strong and simple, and a bond develops with the land.

Two nights ago, after dinner at our campsite on Two Ocean Plateau, we took a short stroll across the ridge. The sun had already set and while we walked its last pink glow drained from the sky. Outlines dimmed and by the time we turned back toward camp the evening had become too dark for us to see. We stumbled on through the blackness, over rocks and fallen timber, following the glint of our campfire through the trees. And somehow this fire in the wood seemed to offer more warmth and protection than any domestic hearth. We felt as we neared it that we were coming home.

We ride on, past a tiny creek pungent with sulfur, its water the temperature of hot soup. Strands of green and blond algae, resembling hair, flow along the bottom. Wild peppermint grows on the trail, and as our horses' hoofs crush the plants their essence mingles strangely with the sulfur fumes from the creek.

The trail leads on, following the Snake River into ever-broadening meadows, the landscape rolling ever more gently, until far ahead we hear a low, complaining moan. It is the sound of automobiles on the highway. The speed limit in Yellowstone is 45 miles per hour, but to us these cars seem to be moving at breakneck speed. A family of tourists stands beside an automobile at the highway's edge. They see, coming toward them across the Snake River bottomland, a string of six horses and four dusty, unshaven riders in funny-looking Western hats. The sight is as curious, I imagine, as anything they have seen all day.

A Huge Cauldron on the Boil

PHOTOGRAPHS BY RICHARD PHILIPS

Some of the most theatrical displays of nature in rampaging action take place every day at Yellowstone National Park. In a never-ceasing spectacle, jets of boiling water and steam spurt from some 200 geysers while thousands of hot springs, mud pots and fumaroles hiss, bubble and steam without pause.

The show has been running for about 600,000 years. It began with a climactic volcanic eruption that spewed lava over most of what is now the park area. The molten rock, or magma, that incited the outburst still lies below, in some places only 11,000 feet beneath the surface. It is hot enough—perhaps 2,000°F.—to keep Yellowstone's cauldron in constant boiling ferment.

The action quickens when water from Yellowstone's snow and rain —some 25 inches of precipitation a year—seeps down deep enough through the porous lava rock to be heated by the magma. The steam that forms seeks a way up and out. If unobstructed, it puffs out through holes in the ground called fumaroles. If the nearby soil is wet, the result is gurgling, bubbling mud.

But it can be difficult for steam to reach the surface. Much of subterranean Yellowstone is a maze of cracks and fissures, an underground plumbing system full of water from the surrounding porous rock. Because this volume of water weighs so much, the heated water at the bottom of the system lacks room to expand and boil. It keeps hotting up, to a degree perhaps triple that of its normal boiling point. Finally the pent-up power below equals that of the water above, producing a standoff between two enormous forces: steam energy and water pressure.

Now, where the "pipes" of the plumbing are wide enough, the heated water and the steam bubbles move up unchallenged, emerging at ground level in gently welling hot springs. But if the pipes are constricted, the equilibrium between water pressure and steam energy is shattered, and the phenomenon of the geyser—a word derived from the Icelandic *geysa,* to gush—is unleashed. Unable to rise freely, the expanding steam bubbles begin to push the water ahead of them. Some of it splashes out at the surface, reducing the pressure and thus permitting more steam energy to be loosed, more water ejected. The climax is an eruption in which torrents of water followed by clouds of steam spew from the geyser's mouth.

Yellowstone's most dependable geyser, Old Faithful, caps an eruption by shooting a plume of steam from its underground plumbing system. The plumes sometimes rise 200 feet and the vapor cloud is most dense when the air is chilly, as on this occasion.

Inch-high terraced steps of sinter cover a slope near Giantess Geyser. They develop where the ground's gradual incline slows the runoff of geyser water so that it evaporates slowly and leaves its burden of minerals behind.

A puddle in Biscuit Basin harbors the stone "cookies" (left) for which the place is named. Two to three inches across, these forms may join the surrounding sinter if not broken by rare upheavals in their calm pools.

Fragile Formations in Stone

Yellowstone's geysers carry stupendous volumes of water to the surface. Giant Geyser, for example, has been known to disgorge a million gallons in a single 90-minute blow-up. Some of the water seeps back into the ground. A great deal more runs off on the Atlantic-drainage side of the Great Divide, by way of the Firehole River, a tributary of the Madison River, and eventually by way of the Missouri and Mississippi rivers. But thousands of gallons of water evaporate on the surface where they emerge, and the minerals with which they are laden harden into rock deposits most commonly consisting of sinter, or solid silica, and a variety of limestone called travertine. So fragile are some of these formations that a footstep would crumble them.

Where the ground slopes gently away from a geyser, as at Giantess, the evaporating runoff builds up miniature terraces of sinter *(upper left)*. Elsewhere, sinter assumes delicate rounded shapes as it crystallizes *(lower left)*; these forms appear around geysers with nearby depressions where puddles accumulate.

Even the water that returns into the ground leaves exotic traces en route. Over the years deposits of travertine from the runoff of many hot springs have built up in odd shapes around the mouth of New Blue Spring Fissure *(right)*, a long narrow crack in the earth's crust at Mammoth Hot Springs.

Mineral-laden water trickling back into New Blue Spring Fissure decorates the foot-wide crevice with travertine nodules and stalactites.

The ghostly stumps of pine trees, enveloped and petrified by deposits of sinter, form the unique eight-foot-high cone of Grotto Geyser.

The Varied Shapes of Geyser Vents

No two geysers act or look alike. Differences in the earth's internal structure and in the amount of water and heat cause wide variations of their eruptions, and this in turn determines each geyser's appearance.

In general, however, a geyser may be categorized in one of two ways: the fountain type or the cone type. The fountain type erupts from the middle of a pool of water that may vary in diameter from only two feet to almost 20 feet, as in the case of Great Fountain Geyser shown on page 156. As the geyser wells up, the pool rises. After the eruption, the pool again drops, sometimes draining completely into the geyser's vent *(lower right)* and always leaving a deposit of sinter at the surface.

Most geysers that lack pools build up sinter cones around their vents. The cones are small *(upper right)* in the case of new geysers or those with a slight water flow, and much larger in the case of older or more active geysers, such as Castle *(page 157)* and White Dome *(page 156)*, whose cones are 15 and 25 feet high respectively. Another cone geyser, Grotto *(left)*, has its own special distinction. At some time in the past it became dormant and pine trees took root around its mouth. Then Grotto reawakened, killed the trees with its steamy breath and covered their stumps with sinter that gradually petrified them and built up around them the cavelike cone for which the geyser is named.

The low-lying cone around the vent of this nameless geyser, one of hundreds of relatively frugal bubblers at Yellowstone, indicates that it has been in action for only a short time, perhaps no more than 20 years.

At the end of a fountainous eruption, the pool of water that most of the time completely covers the vent of Echinus Geyser drains back into it, exposing the sinter deposits that form a plating on the surrounding rock.

Old Faithful shoots a column of boiling water some 130 feet into the air; crowning the eruption, a cloud of steam rises 100 feet higher and

drifts slowly downwind to form a graceful white curtain across Upper Geyser Basin.

The Geyser That Never Fails

The star performer among all the thermal attractions of Yellowstone Park is Old Faithful. It was named and first described by an exploring party, led by Henry Washburn, that camped near the geyser in 1870. Since then it has become unquestionably the most popular spectacle in the park and probably the best-known geyser in the world.

Yellowstone's geysers are many and varied *(overleaf)*, and some of them outdo Old Faithful in certain ways. It is not, for example, the largest of the park's geysers; though its waterspouts often soar to 180 feet, they cannot match Steamboat Geyser's 300-foot-high eruptions. Nor are Old Faithful's five-minute-long outbursts the most protracted; Giantess Geyser sometimes erupts for as long as a day and a half. Plume Geyser has a more regular schedule; it goes off like clockwork every 25 minutes, while the intervals between Old Faithful's eruptions range from 33 to 148 minutes.

Old Faithful's fame is based, instead, on its remarkable constancy. Other geysers have become dormant or revised their schedule of appearances, but Old Faithful never disappoints. Winter and summer, without fail, on the average of 22 times a day, Old Faithful sends 10,000 to 12,000 gallons of boiling water and steam surging skyward. Since its discovery by the Washburn party it has erupted more than 820,000 times and shows no sign of slackening.

CLIFF GEYSER

WHITE DOME GEYSER

AFRICA GEYSER

GREAT FOUNTAIN GEYSER

BEEHIVE GEYSER

CLEPSYDRA GEYSER

PINK CONE GEYSER

CASTLE GEYSER

6/ Glacier's Many Faces

Give a month at least to this precious reserve. The time will not be taken from the sum of your life...it will indefinitely lengthen it and make you truly immortal. JOHN MUIR/ *OUR NATIONAL PARKS*

As the road spins northwest through the Montana plains toward Glacier National Park, where the Continental Divide crosses the border into Canada, the country changes. The differences are subtle: a coolness in the air, a softening of the land into pleasantly rolling hills, a welcome humidity. In the wheat fields along the way, grain grows taller than in the plains farther south. The hillsides change, gradually but surely, from brown to green.

Glacier Park is unlike any other section of the United States Rockies. The mountains, though not as high above sea level as the massive pyramids of central Colorado, rise even more abruptly from the plains. Their tops, refined by ice and snow into elegant needle-sharp horns and minarets, possess a craggy beauty distinctly their own. Elsewhere in the Rockies the mountains impress with sheer bulk; like adolescents who have grown big too fast, they often seem ungainly. Glacier's summits reach up with the grace and proportion of brush strokes in a Japanese scroll. While other ranges display an occasional glacier, here patches of glacial ice sparkle along the high ridges like silver confetti. Even the sky is different: softer, moister, with great shifting clouds that trail off imperceptibly into the surrounding blue.

The aspect of the Great Divide too is unique in Glacier. Nowhere else does the barricade rise with such decisive and uncompromising suddenness. For the 55 miles or so of the park's length, the divide threads the narrow edge of a series of rock walls that, in most places, jut straight up more than half a mile.

I approached the park from the east, and entered it at Sherburne Lake, a finger of water that pokes between two steep promontories into a deep glacial valley. Suddenly I was surrounded by mountain peaks. The effect was overpowering. Sherburne Lake lies 4,774 feet above sea level. The stubbiest mountain next to it, Altyn Peak, tops out at 7,900 feet. Others soar well above 9,000 feet, or nearly a mile higher than the lake. There is almost no transition. A green apron of spruce and aspen climbs several hundred feet above the valley floor, and then quits. Above it the mountains rise with such perpendicular abruptness that no tree can root on them. They become sheer rock walls, gray as slate, relieved by glittering white pockets of snow and glacial ice.

My home base was a rambling wooden structure just beyond Sherburne Lake called Many Glacier Hotel. Water bounds the hotel on two sides: Sherburne Lake to the east behind a low ridge, and 30 yards to the west Swiftcurrent Lake, also apparently hemmed in by mountains on all sides. This is an illusion. The valley turns a corner and continues west, where other lakes nestle. Above Swiftcurrent Lake is another pool of water called Lake Josephine, and above that, hidden among the peaks, Grinnell Lake.

The source of Grinnell Lake is a blanket of ice 300 acres in extent called Grinnell Glacier. From the balcony in front of my room at Many Glacier Hotel I could just make out the sunlight glinting off its surface. Behind the glacier rose a headwall of awesome steepness that culminated in a filigree of rock spires. This stone barricade is known as the Garden Wall, and along its top runs the Great Divide.

I could see only a small stretch of the Garden Wall from the hotel. An intervening mountain obstructed the view. But I could follow the divide as it emerged on the mountain's other side along an identical rampart, the Pinnacle Wall, equally jagged and with its own drapery of ice and snow. I resolved to go there for a closer look.

A horse trail leads to the Pinnacle Wall, culminating at a glacial tarn called Iceberg Lake. I set out the next morning, and reached the lake soon after midday. I tethered my horse to a hitching post and hiked the last 200 yards to the lake shore. The path led through a hurly-burly of alpine flowers. The growing season is so short here near the snow that everything blooms at once. Summer plants—paintbrushes, yellow arnica, heather with its tiny pink bells—mingled with the spring beauties and glacier lilies of early spring.

Winter still gripped the lake itself. Chunks of ice clotted its surface, and a solid drift of snow extended down the headwall opposite and out into the lake. Not too many years ago this white cascade had probably been a glacier. But during the past century the park has grown warmer, and the snow at Iceberg Lake has dwindled to the point where it no longer compacts into glacial ice. It is simply a large snowbank.

All through the park, in fact, the glaciers have been melting away. In the 1880s, the peak decade of the present ice accumulations, Grinnell Glacier spread more than 500 acres down the east side of the Garden Wall. It has retreated so alarmingly that today a corridor of rock cuts across its midriff, splitting it in half. Since the 1950s, however, the ice has been shrinking at an ever-decreasing rate. Some geologists even speculate, with what is an admittedly partisan eagerness, that a new ice age may be on its way.

In any case, the ebb and flow of today's ice is a mere jiggle compared with the prodigious movements of earlier ice sheets that scoured Glacier Park. At periodic intervals in the past, when the climate of North America was much colder, the entire park has been enshrouded in a solid canopy of ice, with only a few isolated mountain crags poking through. The last big freeze began about 70,000 years ago, with the onset of a major chill known as the Wisconsin Ice Age. It continued, with occasional thaws, until about 8000 B.C., when most of the ice melted. Since then, the park's landscape has not changed very much, because today's glaciers, which started accumulating about 300 years ago, are too small to do much major surgery on the rock.

But what an awesome architecture those ancient glaciers left behind. The cirque that cradles Iceberg Lake looms overhead 3,000 feet, with a steepness that seems to defy the pull of gravity. The cirque is large, about a mile from rim to rim, but its effect is almost claustrophobic. It closes around the lake on three sides, like a cave with the roof blown off. To the left rises the vertical escarpment of a 9,293-foot mountain, Mt. Wilber, its summit sharpened to the spear point of a perfect glacial horn. So precipitous is this particular crag that no one was able to reach its top until 1923, despite a number of attempts.

Men may have trouble with this terrain, but there is one creature in Glacier Park that has no difficulty with it at all—the mountain goat. He is a shaggy white animal of deceptively boxlike proportions, with a large rectangular torso perched on four short legs, a small head with two ebony horns and an expression of perpetual bewilderment on his face. The mountain goat's bumbling appearance is all sham, however. He lives primarily above timberline, summer and winter alike, apparently oblivious to cold, wind, blizzard and the continual possibility of a misstep off a mountainside. He can leap with total self-assurance along cliff faces, up precipices and over chasms that would terrify the steeliest mountaineer. His secret is anatomy. A mountain goat's torso, though a clumsy rectangle when viewed from the side, is actually quite narrow across the shoulders, a conformation that allows him to move easily along narrow trails on a mountain face. In addition, each foot has a unique no-skid construction. A spongy pad protrudes from the center of each hoof, giving roughly the same kind of traction provided by a rubber snow tire with steel spikes.

Iceberg Lake is a notoriously good spot for goat watching. I was assured of this by one of the park rangers who regularly conducts nature walks there. The ranger is so sure of finding goats that he promises his hiking party—in advance—that he will buy everyone ice-cream cones if the animals do not appear. He has never had to pay up, though once he came close. The afternoon was a particularly hot one, the kind of day when nothing moves, and by the time the party had sweated its way to Iceberg Lake the cirque was deserted. The ranger had already begun to add up the cost of 25 ice-cream cones when he realized that because of the heat, all the goats must be cooling their heels behind the snowbanks midway up the cirque walls. So he started climbing. It took him a hot, wearisome half hour to reach the snow, but the climb bore results. A string of six goats came trotting out from behind a snowbank, to the applause of the crowd below.

No such exertion was necessary in my case. I trained my glasses on the edge of a snowbank, and sure enough a tiny white rectangle stood up and started walking. I soon discovered a dozen similar white spots. They were all high up, and so far away I could only just distinguish them as mountain goats. One glance at the grim escarpments overhead convinced me that I did not really want to climb any closer.

Before coming to Glacier Park, my plan had been to hike across one of the glaciers, scramble up the headwall behind, and cross over the bottom of the Great Divide to the west slope. Now, from the bottom cirque at Iceberg Lake, I could see that such a procedure would be impossible; I am no mountain goat. I would have to be satisfied with a more prosaic approach, by automobile. So the next day I set out on the highway over Logan Pass to the south, crossing the divide from east to west. The trip revealed a fascinating aspect of the divide in Glacier Park. Here the divide, to a particularly noticeable degree, separates not only watersheds, but climates. All through the mountains, east slopes tend to be drier than west slopes. But in Glacier the contrast is startling.

Sure-footed white mountain goats, three adults and a kid, gather at a mineral lick on a craggy height of the northern Rockies. The lick is a natural outcropping of rock rich in the salt the goats' diet apparently must have.

I approached Logan Pass through the same panoramic countryside that had led me to Many Glacier Hotel—lake-filled valleys, a crazy quilt of spruce and aspen, abrupt and jagged mountains chewing into the sky like sharks' teeth. The highway bore west into the mountains and started zigzagging toward timberline. As the land became more vertical the vegetation thinned, so that large stretches of stony brown soil showed through. The road made a final switchback, reached the pass and crossed the divide. There, abruptly as a door opening, another landscape came into view. It was as though a symphony orchestra, in the middle of a musical phrase, had suddenly switched keys.

All of northwestern Montana is a bit moister than the rest of the Rocky Mountains. At my lodge on the east slope of the park I had been impressed by the greenness of the land. But now, as I looked into the valleys west of Logan Pass, the country seemed as green as Ireland, as wooded as the coast of Maine. Forest did not merely cover the land, it overwhelmed it. Elsewhere in the Rockies, even in the darkest spruce groves, I had always been conscious of the rock skeleton underneath. Now, even above timberline, vegetation almost obscured the crags. For several miles below the pass the highway ran along the west side of the Garden Wall, the same bleak ridge I had noticed from my lodge. But here its steep slopes were mantled with plants. Flowers drooped down the bank in long shaggy fronds, a verdant waterfall of aster, fleabane, spirea, wild hollyhock.

Moisture trickled through the flowers from invisible springs and snowbanks higher up the wall, and poured out onto the road. And moisture was the reason for this green profusion. In crossing the divide I had moved into another climate zone, at once more watery and more temperate. So marked is the difference in climate that the regional newspaper, the Great Falls *Tribune,* publishes two weather reports, one for each side of the divide.

The winds are generally strongest and the temperatures harshest east of the divide, in the lee of the mountains. This odd state of affairs occurs, in part, because the prevailing westerlies change speed as they ride over the mountains. They climb laboriously up the western slopes, dropping their moisture along the way. But once over the top they may start careering downhill like a roller coaster gone berserk. One such

downhill air current that has a certain notoriety in Glacier, and along the entire lee slope of the Rocky Mountains, is the chinook. It comes only in winter, and its effect is usually benign, for it brings in a sudden spell of hot, dry weather that can melt a foot of snow in a matter of hours. (The Blackfoot Indians called it snow eater.) But occasionally the chinook arrives with such headlong speed that it packs the destructive power of a small hurricane. One particularly ferocious variation, which local weathermen call a Klondike chinook, blows down from Alaska. This one is cold, and so does not even melt the snow. On January 16, 1972, a Klondike chinook smashed into Browning, Montana, about 15 miles east of Glacier Park. It hit a freight train rolling along an unprotected stretch of track, and derailed seven flatcars, flipping a couple of them over an embankment.

The wind never blows as hard west of the divide, nor do temperatures shift so wildly. The climate is moist and temperate to a degree unusual in the Rockies. In McDonald Valley, the verdant basin I looked down upon from the road west of Logan Pass, clouds cover the sky almost 250 days out of the year. This is not typical Rocky Mountain weather. It belongs by rights to country 500 miles away, on the edge of the continent. McDonald Valley is the easternmost outpost of a mild but soggy climate system that prevails in the Pacific Northwest.

As I followed the highway into the valley, I might have been driving through the great Pacific evergreen forests of Oregon or Washington. At one point I parked the car and hiked through a grove of giant red cedars. It was as though someone had planted a greenhouse in the nave of a church. Cedar trunks rose in lofty pillars to a green vault 90 feet overhead. So little light seeped through that I had to wait until my eyes adjusted before I could see the path. I became aware of a sensation I had missed during my entire odyssey along the Great Divide. I could smell the earth—rich, damp, loamy, an odor compounded of ferns, mushrooms, decaying logs. And as I walked I noticed something else: my boots raised no dust. Almost every trail in the mountains leads across a dry and stony soil that sends up miniature dust flurries when walked upon. I had become so accustomed to feeling pebbles through my boot soles that the sensation no longer registered in my brain. Here not a pebble, nor even a patch of bare dirt. A green upholstery of moss covered every inch of ground. My boots sank into it as though I were walking across an innerspring mattress.

The contrast with Iceberg Lake was extraordinary. Here in the sedate and gentle world of the cedar forest, it was hard to imagine that only a dozen miles away, on the east slope of the Great Divide, the land could be so harsh and unrelenting. I returned to my car and drove back east over Logan Pass to my hotel, back to a country of stark precipices, impassable rock walls and mountains scraped raw by ice and weather. Near the hotel I noticed several stands of aspen. Their tops were all cut down to exactly the same level, as though they had been sheared off by a pair of giant hedge clippers: the effect of winter windstorms blowing down the lee slope of the divide.

My last day in the Rocky Mountains dawned overcast and stormy. Soupy gray clouds rolled down the mountains surrounding Swiftcurrent Lake, obscuring the peaks; a heavy mist smoked up from the lake to meet them. After breakfast I hiked out into the drizzle, trailing south along a ridge of buff-colored rimrock behind the hotel. The rock was slippery in the dampness, its surface ground smooth by the ancient glacier that had hollowed out the Swiftcurrent Valley.

I followed the ridge into a margin of timber. Mist swamped the hotel and its outbuildings, shielding them from view. I had hiked perhaps half a mile; it might have been a hundred for all of civilization I could see. Two gray birds with black and white wings, Clark's nutcrackers, perched on a wind-stripped tree, its gray branches dead as driftwood. The nutcrackers squawked, a harsh and guttural cry, then flew into the timber, their tail feathers fanning out into black and white stripes, formal as evening clothes. A pair of chipmunks clowned among the rocks at my feet. They ran about with great vitality and fuss, collecting nuts and seeds against the approaching winter.

I did not stay long. The clouds darkened overhead, and the drizzle hardened into rain. But to the west, across Swiftcurrent Lake, the weather seemed to clear a bit. The clouds thinned and the outlines of the mountains began to appear—first their flanks, then the near peaks, condensing out of the gray swirl. For a brief moment the clouds pulled back from the Pinnacle Wall, unveiling the fine, jagged crest of the divide itself. So stern and upright did the barrier seem that I could not believe any man had breached it or that I had walked it myself in the mountains farther south. Then the curtain dropped again and the Great Divide dissolved from sight.

September in the Park

PHOTOGRAPHS BY DAN BUDNIK

When Dan Budnik went to Glacier National Park to photograph the portfolio that follows, he fulfilled a long-held desire to experience at first hand a landscape that had impressed him, solely from his boyhood reading about it, as one of the most magnificent to be found anywhere on the continent.

As it turned out, his experience embraced not only a scenic feast but a quick education in the vagaries of weather in mountain terrain. Summer comes late to the high places of Glacier Park, and winter comes early and hard. Within the brief span of a 10-day visit early in September, Budnik witnessed a dazzling variety of weather changes that seemed to telescope summer, autumn and winter: from warm, brilliant sunshine and temperatures in the 70s to chilling thunderstorms to hailstones and a sudden snowfall.

Budnik conducted his initial exploration from a light plane in the late afternoon of a crystal-clear day. "I was fully prepared for spectacular mountains," he says, "but not for the overwhelming vastness of the park. From the air, the access roads faded, what few other signs of man's presence there were had disappeared completely, and seemingly trackless wilderness extended in every direction." The slanting rays of the setting sun highlighted the many finger lakes carved by the glaciers, and lent a deceptive warmth and gentleness to the contours of the formidable glaciers themselves.

After his aerial survey Budnik took to the park's trails for a few days of rugged backpacking. Another, smaller world came into view, a world of subtle details that particularized the charms of the park: blooming wild flowers caught by the abrupt advent of snow, a thin shaft of sunlight slicing through the mist, little patches of melting snow glistening in the morning sun, twisted pieces of driftwood half submerged in a shallow lake.

Budnik came away from Glacier tempted to cherish it as a private, personal retreat, yet at the same time eager to share his appreciation of the place with others. But he warns that getting to know it is no easy matter. "The best way to explore Glacier," he says, "is to carry a pack on your back and go for the high ground. Only then can you get away from the cars and their safety-belted sightseers. You can spend a whole summer discovering the mountains and never use a paved road."

LATE AFTERNOON OVER SPERRY GLACIER

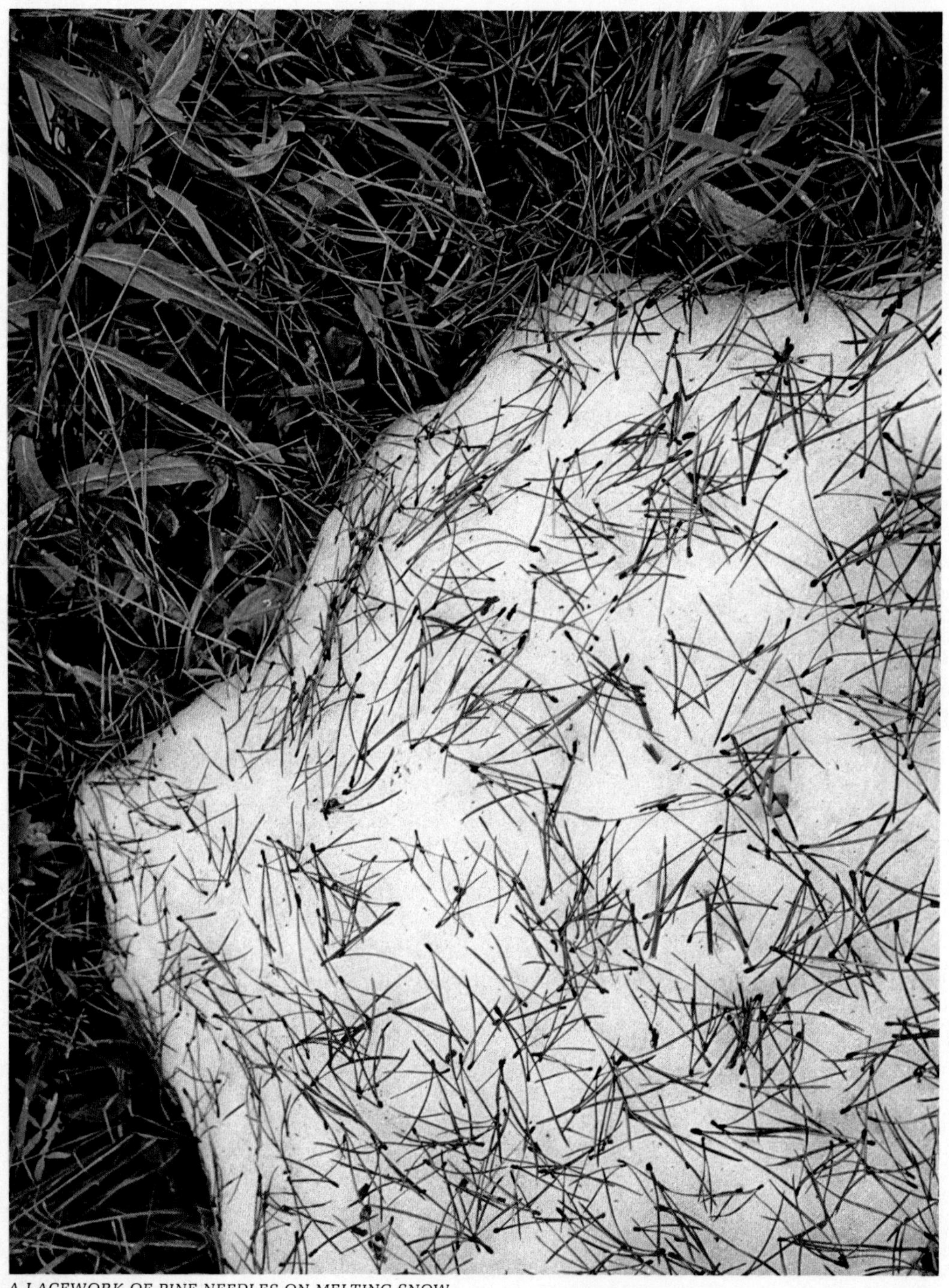

A LACEWORK OF PINE NEEDLES ON MELTING SNOW

LODGEPOLE PINES MASSED BELOW BEAR MOUNTAIN

SAINT MARY LAKE CRADLED BY CLOUD-CAPPED PEAKS

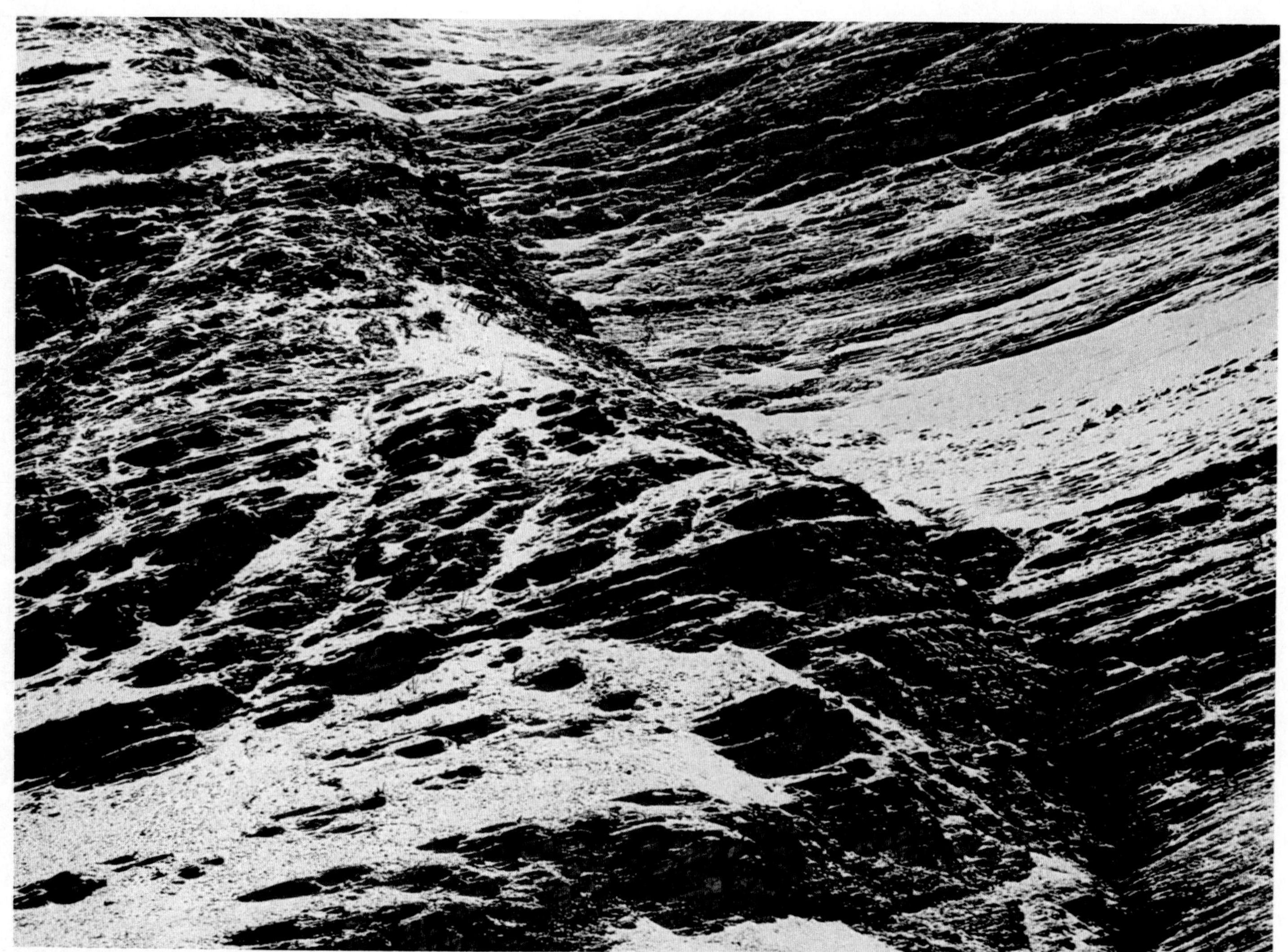

A DUSTING OF SNOW ON A STARK ROCK FACE

INDIAN PAINTBRUSH AND YARROW DEFYING A WINTRY WARNING

A SNOWFALL ON FISHERCAP LAKE

MORNING MIST AND SUN ON FIR TREES

ASPEN TRUNKS LONG AGO DEFORMED BY WINTER SNOWS

THE ENSHROUDED PYRAMID OF MATAHPI PEAK

Bibliography

*Also available in paperback.
†Available only in paperback.

*Baker, William H., and Earl J. Larrison, *Wildlife of the Northern Rocky Mountains.* Naturegraph Company, 1961.

Bartlett, Richard A., *Great Surveys of the American West.* University of Oklahoma Press, 1962.

Bauer, Erwin A., *Treasury of Big Game Animals.* Harper & Row, 1972.

Chittenden, Hiram Martin, *The Yellowstone National Park.* University of Oklahoma Press, 1964.

†Chronic, John and Halka, *Prairie, Peak and Plateau; a Guide to the Geology of Colorado.* Colorado Geological Survey, 1972.

Cook, Charles W., David E. Folsom, and William Peterson, *The Valley of the Upper Yellowstone.* University of Oklahoma Press, 1965.

Craighead, John J. and Frank C., Jr., and Ray J. Davis, *A Field Guide to Rocky Mountain Wildflowers.* Houghton Mifflin Company, 1963.

Davis, Clyde Brion, *The Arkansas.* Farrar & Rinehart, 1940.

†Dyson, James L., *The Geologic Story of Glacier National Park.* Glacier Natural History Association, 1969.

†Dyson, James L., *Glaciers and Glaciation in Glacier National Park.* Glacier Natural History Association, 1966.

†Eberhart, Perry, and Philip Schmuck, *The Fourteeners: Colorado's Great Mountains.* The Swallow Press Inc., 1970.

†Keefer, William R., *The Geologic Story of Yellowstone National Park.* U.S. Geological Survey, 1971.

†Kirk, Ruth, *Exploring Yellowstone.* University of Washington Press, 1972.

†Langford, Nathaniel Pitt, *The Discovery of Yellowstone Park.* University of Nebraska Press, 1972.

Lavender, David, *The Big Divide.* Doubleday, 1948.

Lavender, David, *The Rockies.* Harper & Row, 1968.

Lavender, David, *Westward Vision.* McGraw-Hill Book Company, Inc., 1963.

Lewis, Meriwether, and William Clark, *The Journals of Lewis and Clark,* edited by Bernard DeVoto. Houghton Mifflin Company, 1953.

†Lystrup, Herbert T., *Hamilton's Guide to Yellowstone National Park.* Hamilton Stores Inc., 1971.

†McDougall, W. B., and Herma A. Baggley, *Plants of Yellowstone National Park.* Yellowstone Library and Museum Association, 1956.

†Morgan, Dale L., *Jedediah Smith and the Opening of the West.* University of Nebraska Press, 1953.

Murie, Olaus J., *The Elk of North America.* The Stackpole Company, 1951.

†Nelson, Ruth Ashton, *Handbook of Rocky Mountain Plants.* Dale Stuart King, 1969.

Ormes, Robert, *Guide to the Colorado Mountains,* 6th ed. The Swallow Press Inc., 1970.

Pearl, Richard M., *Colorado Gem Trails,* 3rd ed. The Swallow Press Inc., 1972.

Ryden, Hope, *America's Last Wild Horses.* E. P. Dutton & Co., Inc., 1971.

†Scharff, Robert, ed., *Glacier National Park and Waterton Lakes National Park.* David McKay Company, Inc., 1967.

†Scharff, Robert, ed., *Yellowstone and Grand Teton National Parks.* David McKay Company, Inc., 1967.

Sprague, Marshall, *The Great Gates, the Story of the Rocky Mountain Passes.* Little, Brown and Company, 1964.

Stuart, Robert, *The Discovery of the Oregon Trail.* Charles Scribner's Sons, 1935.

Sutton, Ann and Myron, *Yellowstone, a Century of the Wilderness Idea.* The Macmillan Company, 1972.

†U.S. Geological Survey, *Geysers.* U.S. Government Printing Office, 1971.

Zwinger, Ann H., and Beatrice E. Willard, *Land above the Trees; a Guide to American Alpine Tundra.* Harper & Row, 1972.

Acknowledgments

The author and editors of this book are particularly indebted to Jack D. Ives, Director, Institute of Arctic and Alpine Research, University of Colorado, Boulder. They also wish to thank the following persons and institutions in Colorado: Richard L. Armstrong, Institute of Arctic and Alpine Research, San Juan Avalanche Project, Silverton; Roger G. Barry, Institute of Arctic and Alpine Research, University of Colorado, Boulder; Roger C. Brown, Vail; David Bucknam, Littleton; Colorado Mountain Club, Denver; James H. Davis, Denver Public Library; John Davis, Center; Blake Dunn, Allemosa; Robert Gnam, U.S. Forest Service, Denver; Henry Langsford, National Center for Atmospheric Research, Boulder; Duane G. Lankford, Denver; Clifton Merritt, Wilderness Society, Denver; Richard Munro, Denver; Jean Naumann, Vail; Herbert Naumann, Colorado Division of Highways, Vail; Robert Ormes, Colorado Springs; Marshall Sprague, Colorado Springs; David Sumner, Morrison; Jack Washichek, U.S. Soil Conservation Service, Denver; Patrick J. Webber, Institute of Arctic and Alpine Research, University of Colorado, Boulder. In Wyoming: Jack Anderson, Superintendent, Yellowstone National Park; Edmund Bucknell, Yellowstone National Park; William I. Crump, Lander; Don Davis, Lander; Rod Hutcheson, Yellowstone National Park; Vincent R. Lee, Jackson; James Moorhouse, U.S. Bureau of Land Management, Cheyenne; Ken and Ann Richardson, Lander; John W. Stockert, Yellowstone National Park; Kurt Topham, Yellowstone National Park; John F. Turner, Moose; Bob Woodie, Yellowstone National Park. Also William J. Briggle, Superintendent, Glacier National Park, Montana; Donald F. Bruning, Assistant Curator of Ornithology, New York Zoological Park, New York City; David Casteel, Glacier National Park, Montana; Grayson Cordell, Meteorologist in Charge, National Weather Service, Helena, Montana; Joseph A. Davis, Scientific Assistant to the Director, New York Zoological Park, New York City; Jerome S. De Santo, Glacier National Park, Montana; Sidney S. Horenstein, Department of Invertebrate Paleontology, The American Museum of Natural History, New York City; Clifford Martinka, Research Biologist, Glacier National Park, Montana; Larry G. Pardue, Plant Information Specialist, New York Botanical Garden, New York City; Hope Ryden, New York City.

Picture Credits

Sources for the pictures in this book are shown below. Credits for the pictures from left to right are separated by commas; from top to bottom they are separated by dashes.

Cover—Dan Budnik. Front end papers 2, 3—Dan Budnik. Front end paper 4, page 1—Bob Waterman. 2, 3—Bob Gunning. 4, 5—Dan Budnik. 6, 7—B. Riley McClelland. 8, 9—David Sumner. 10, 11—Dan Kramer. 12, 13—Ed Cooper. 18, 19—Maps by R. R. Donnelley Cartographic Services. 24, 25—Dan Budnik. 26—Bob Waterman. 33—Richard L. Armstrong. 37 through 45—Dan Budnik. 50—International Museum of Photography at George Eastman House. 54, 55—Robert Walch. 60—Map by Vincent Lewis. 60 through 71—David Cavagnaro. 76, 77—Tom Tracy. 80—C. A. Morgan. 84, 85—Bob Waterman. 89—Judy Sumner—Betty Randall, Betty Randall, Betty Randall—Robin Way. 92—Betty Randall. 97 through 101—Bob Waterman. 104, 105—Dan Kramer. 109—Dan Kramer. 112, 113—Dan Kramer. 116, 117—Dan Kramer. 121—Frank Lerner courtesy The New York Public Library, Astor, Lenox and Tilden Foundations except top right Stewart Cassidy. 122, 123—Montana Historical Society except top left James K. Morgan. 124, 125—Montana Historical Society except top right Maurice G. Hornocker. 126, 127—Denver Public Library, Western History Department except top left C. A. Morgan. 130—Harald Sund. 132, 133—Harald Sund. 134—Bill Browning. 138, 139—Harald Sund. 142, 143—Harald Sund. 149 through 157—Richard Philips. 163—C. A. Morgan. 167 through 179—Dan Budnik.

Index

Numerals in italics indicate a photograph or drawing of the subject mentioned.

Printed in U.S.A.